Sifang Li
Manufacture of Fine Chemicals from Acetylene

Also of Interest

High Pressure Organic Synthesis
Davor Margetic, 2019
ISBN 978-3-11-055595-0, e-ISBN 978-3-11-055684-1

Organic Chemistry.
Fundamentals and Concepts
John M. McIntosh, 2018
ISBN 978-3-11-056512-6, e-ISBN 978-3-11-056514-0

Organophosphorus Chemistry.
Novel Developments
György Keglevich (Ed.), 2018
ISBN 978-3-11-053453-5, e-ISBN 978-3-11-053583-9

Industrial Organic Chemistry
Mark Anthony Benvenuto, 2017
ISBN 978-3-11-049446-4, e-ISBN 978-3-11-049447-1

Sifang Li

Manufacture of Fine Chemicals from Acetylene

DE GRUYTER

厦门大学出版社 XIAMEN UNIVERSITY PRESS | 国家一级出版社 全国百佳图书出版单位

Author
Prof. Sifang Li
College of Chemistry and Chemical Engineering
Xiamen University
422 Si Ming South Road
361005 Xiamen
China
sfli@xmu.edu.cn

ISBN 978-3-11-071496-8
e-ISBN (PDF) 978-3-11-071499-9
e-ISBN (EPUB) 978-3-11-071501-9

Library of Congress Control Number: 2021933771

Bibliographic information published by the Deutsche Nationalbibliothek
The Deutsche Nationalbibliothek lists this publication in the Deutsche Nationalbibliografie; detailed bibliographic data are available on the Internet at http://dnb.dnb.de.

Cover Image: shank_ali / E+ / Getty Images
Typesetting: Integra Software Services Pvt. Ltd.
Printing and binding: CPI books GmbH, Leck

www.degruyter.com

Contents

Chapter 1
Introduction

1.1 Overview of fine chemicals from acetylene

China is short of petroleum and natural gas resources but rich in coal resources, so it is an inevitable trend to further develop energy and chemical products with coal as the source. With the development of new technology for manufacture of acetylene from coal by plasma pyrolysis, acetylene may become a cheaper chemical raw material. Therefore, the acetylene chemical technology with acetylene as raw material will become an important development direction in the post-petrochemical era and will have a good development prospect.

Acetylene contains relatively active unsaturated triple bonds and can react with many substances. Many scientists around the world have studied acetylene chemistry, especially the German scientist Reppe, who conducted detailed studies on many reactions of acetylene, forming the famous Reppe process and promoting the development of many fine chemicals from acetylene. Other scholars, such as Kucelov, Favorskii, Zelinsky and Newland, also conducted in-depth researches on acetylene chemistry. Starting from acetylene, thousands of organic chemical products can be synthesized, making acetylene a basic raw material for synthetic plastics, synthetic rubber, synthetic fibers, pharmaceuticals, pesticides, dyes, resins and solvents. Even acetylene was once known as the "mother of organic synthesis industry." Since the 1970s, with the rapid development of petrochemical industry, acetylene as the basic raw material of organic products has been gradually replaced by cheaper ethylene and propylene, and many bulk products originally using acetylene as raw material have gradually changed to ethylene and propylene as raw materials. However, compared with ethylene and propylene, acetylene used as raw material has the advantages of short and mature processes. Therefore, it still occupies a certain proportion in the synthesis of many products, especially in the field of fine chemicals. Acetylene has advantages in the preparation of fine chemicals, and many fine chemicals can only be manufactured by acetylene. The series of fine chemicals downstream acetylene is shown in Figure 1.1.

1.2 Acetylene and its reaction technology

1.2.1 Properties of acetylene

Acetylene, also known as aldehydene, ethyne and ethine, is a colorless gas with ether odor at room temperature and atmospheric pressure. CAS Registry Number: 74-86-2. Molecular formula: C_2H_2. Molecular weight: 26.0373. Chemical structure:

https://doi.org/10.1515/9783110714999-001

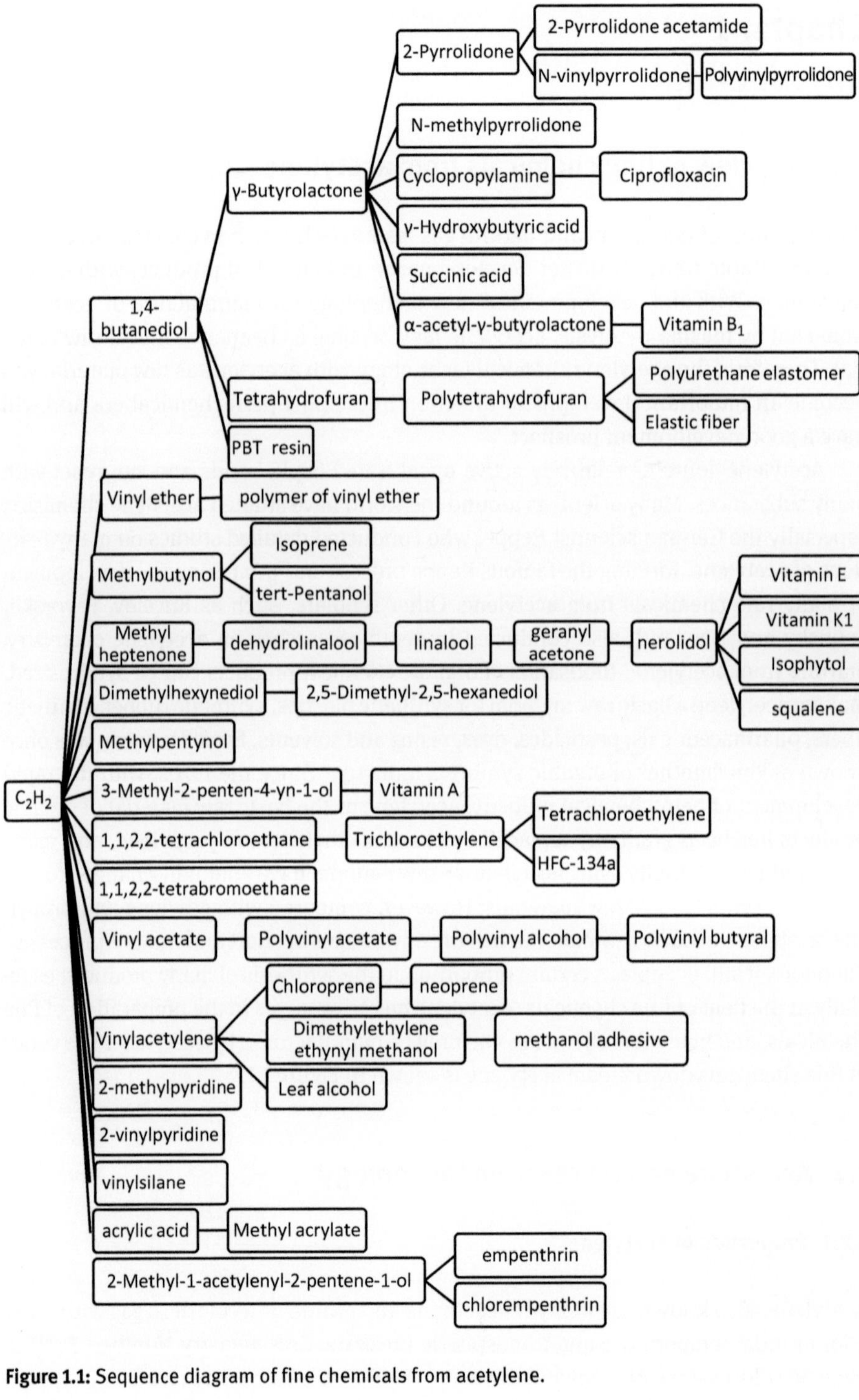

Figure 1.1: Sequence diagram of fine chemicals from acetylene.

$$HC \equiv CH$$

The physical properties of acetylene are listed in Table 1.1. Acetylene is soluble in water and ethanol and very soluble in acetone. The solubility of acetylene in some liquids is shown in Table 1.2. It can be seen that 1 L of water can dissolve 1.1 L of acetylene at 15 °C and under atmospheric pressure, while 1 L of acetone can dissolve 25 L of acetylene under the same conditions. Under a pressure of 1.2 MPa, 1 L of acetone can dissolve 300 L of acetylene. Since acetylene has a very large solubility in acetone, acetylene is pressed into a special cylinder containing acetone and porous material (such as activated carbon) under a pressure of 1.0–1.2 MPa, which is known as "dissolved acetylene." This dissolved acetylene is safe to use, and convenient for transportation and storage [1–3]. Other excellent solvents for acetylene include liquid ammonia, *N,N*-dimethylformamide, *N*-methylpyrrolidone, methylal, dimethoxyethane, dioxane, tetraethylene glycol dimethyl ether and 2-pyrrolidone. A detailed research on the solubility of acetylene in alcohols and ketones can be found in the literature [4].

Table 1.1: Physical properties of acetylene.

Boiling point/°C	−83.6	Heat capacity/J mol^{-1} K^{-1}	
Melting point/°C	−81.5	298	44.04
Density (20 °C)/kg · m^{-3}	1.091	300	44.17
Viscosity (25 °C)/mPa • s	0.0102	400	50.39
Thermal conductivity (25 °C)/W m^{-1} K^{-1}	0.02135	Critical temperature/°C	35.3
Heat of combustion/kJ mol^{-1}	1,255.6	Critical pressure/MPa	6.39

Table 1.2: Solubility of acetylene in liquids at 15 °C and under atmospheric pressure.

Solvent	**Solubility of acetylene (L L^{-1})**	**Solvent**	**Solubility of acetylene (L L^{-1})**
H_2O	1.10	Benzene	4.00
NaCl-saturated solution	0.60	Gasoline	5.70
$Ca(OH)_2$	0.75	Ethanol	6.00
CS_2	1.00	Ethyl acetate	14.80
CCl_4	2.00	Acetone	25.00

When acetylene comes in contact with water, it can form hydrated crystals composed of one molecule of acetylene and six molecules of water at a specific pressure and temperature, which look like ice and snow. The equilibrium state of the hydrated crystal is determined by temperature and pressure (Table 1.3). The highest temperature of hydrated crystals is 15 °C. Above this temperature, the hydrated crystals cannot exist regardless of the pressure. The presence of hydrated crystals in acetylene may cause blockage in the acetylene pipeline or may cause danger due to static electricity generated by the friction between the acetylene gas and the hydrated crystals.

Table 1.3: Relationship between temperature and pressure of acetylene hydrated crystals in equilibrium.

Pressure (gauge pressure) /atm	6	8	10	12	15	20	25	30
Limit temperature/°C	2	4	6	8	10	12	13.5	15

Acetylene gas is relatively unstable and has the risk of decomposition and explosion under high temperature and pressure. The explosion risk of wet acetylene is less than that of dry acetylene, and it decreases further with the increase in its humidity. When acetylene is dissolved in a liquid, its molecules are separated by solvent molecules, and it is stable and safe to handle at high temperature and pressure as long as it remains in the liquid phase.

The explosion limit of acetylene and oxygen mixture is 2.3–93%, and the explosion limit of acetylene in air is 2.3–81%.

The temperature of oxyacetylene flame produced by acetylene combustion can reach more than 2,800 °C, which can be used to cut and weld metals.

When acetylene comes in contact with salts of copper, silver and mercury in aqueous solution, the precipitates of the corresponding metal acetylene are formed, which are explosive and easy to explode when impacted or heated.

1.2.2 Analysis of factors of acetylene decomposition and explosion and methods for safe use of acetylene

The decomposition of acetylene is caused by external energy and certain substances. Typical factors that cause acetylene decomposition and explosion are heat, pressure, electric arc, static electricity, molten metal wires, copper, silver, rust and metal sawdust. Copper and silver can react with acetylene to form copper acetylide and silver acetylide, respectively. Especially in the presence of ammonia, these substances are very unstable and can cause acetylene to decompose and explode [5–15].

Therefore, the use of copper and silver fittings, components, joints and welds should be avoided during storage, transportation and use of acetylene. All equipment, pipelines and connectors with acetylene passing through must be made of stainless steel or rust-resistant iron.

The first step in the decomposition of acetylene is deflagration. In the deflagration step, the initial flame passes through unburned acetylene at a rate of 10–100 m s^{-1}, and the pressure gradually and continuously rises to reach 8–14 times the initial pressure. Deflagration can generally be prevented or inhibited by using a small-diameter reactor and piping, setting a flame arrestor and diluting acetylene with an inert gas such as nitrogen, carbon dioxide, methane and the like. Explosion occurs when deflagration cannot be suppressed, especially if the vessel is large in diameter or acetylene is transported using a long and large-diameter pipe. The resulting flame or wave front velocity increases rapidly when fresh acetylene is consumed until it reaches supersonic speed (1,800–2,000 m s^{-1}) and a violent explosion occurs. This phenomenon is called knocking (or detonation). With a huge energy release (about 220 kJ g^{-1}), the local pressure can rise to 60 MPa or 200 times the initial pressure. Explosive pressure continually jumps with the wave front, and the explosive force is highly directional, which can blow a small hole in the pressure vessel without affecting fixtures and other ancillary facilities.

Even if there is no other detonating factor, decomposition and explosion will occur when acetylene is exposed to high temperatures. Under a pressure of 0.1 MPa, acetylene begins to decompose at 635 °C, while under 0.2 MPa, the average decomposition temperature is 510–540 °C. The temperature at which acetylene decomposes decreases as the pressure increases. Acetylene can decompose and explode at lower temperatures or pressures when certain substances are present. Table 1.4 lists the lowest temperature at which acetylene decomposes and explodes in the presence of certain substances at 0.4 MPa.

Table 1.4: The lowest temperature at which acetylene decomposes in the presence of certain substances.

Substances	The lowest temperature/°C	Substances	The lowest temperature/°C
Calcium carbide	500	Ferric hydroxide	280–300
Aluminum oxide	490	Iron oxide	280
Copper shavings	460	Copper oxide	240
Activated carbon	400		

Table 1.5 reveals the relationship between decomposition pressure and temperature when acetylene is instantaneously induced by a molten platinum wire (0.5 mm diameter). It can be seen that the molten platinum wire can easily decompose acetylene under lower pressure. Other molten metal wires such as iron, copper and aluminum also have similar phenomena.

Table 1.5: Relationship between decomposition pressure and temperature of acetylene induced by molten platinum wire (0.5 mm).

Temperature/°C	Decomposition pressure/atm
15	1.40–1.60
50	1.40–1.48
100	1.25–1.33
140	1.10
150	1.13
180	1.06

Although acetylene is considered to be a high-energy, flammable, explosive and dangerous gas, it has been safely used in industrial production for more than 80 years. Before and during World War II, a large number of tests showed that if the inner diameter of the pipe is less than 25 mm, acetylene could be safely transported over a long distance under several atmospheric pressures. For the large-diameter pipeline, it was found that the safety of acetylene transportation could be improved by filling several bundles of small-diameter pipeline (10–13 mm) into the large-diameter pipeline.

Pure acetylene gas, if diluted with an inert gas, greatly improves its stability and safety. Typical gases studied for diluting acetylene are hydrogen, carbon monoxide, carbon dioxide, nitrogen, ammonia, methane, ethylene and propane. Ammonia is an excellent stabilizing diluent for acetylene, and is used as an extractant for acetylene and a solvent for the reaction. Carbon monoxide also has very good dilution stability, which is of great importance since the mixture of acetylene and carbon monoxide is a reactant for industrial production of acrylic acid and acrylates. Methane used as a diluent is better than nitrogen. Similarly, propane-diluted acetylene is very stable, which is about twice as effective as nitrogen in preventing the decomposition of acetylene. Nevertheless, nitrogen is usually used as the inert diluent for acetylene [16–18].

1.2.3 Acetylene reaction technology

1.2.3.1 Liquid ammonia–alkali metal hydroxide ethynylation technology

The use of potassium hydroxide catalyst in the active medium liquid ammonia allows the aldehyde or ketone to be efficiently subjected to an ethynylation reaction for production of the corresponding acetylenic alcohol. There are a number of solvents that are capable of dissolving large amounts of acetylene, but in view of the solubility, cocatalyst activity, ease of recovery, ease of acetylenic product separation and cost, liquid ammonia has the best comprehensive performance. For example, in 500 mL of liquid ammonia, 6–18 mol of ketone or aldehyde can be converted into the corresponding alkynols in the presence of 1.5 mol of KOH, and the conversion of ketone or aldehyde is in the range of 50–100%. Liquid ammonia–alkali metal hydroxide ethynylation technology has been widely used in the industry. Various ethynylation processes are involved in the industrial manufacture of vitamins A and E, and some perfumeries, in which large amounts of liquid ammonia are utilized and recycled. Table 1.6 lists the conversions for the formation of alkynols from various liquid carbonyl compounds by the liquid ammonia–alkali metal hydroxide ethynylation technology.

The process is operated at a temperature of 20–40 °C and under a pressure of 0.69–2.1 MPa using at least 1.5–2 times excess acetylene to avoid formation of ketone condensation by-products. The amount of liquid ammonia used in the reaction system influences the conversion of the carbonyl compounds, and a decrease in the amount of liquid ammonia reduces the conversion.

Table 1.6: Ethynylation of carbonyl compounds in liquid ammonia–KOH system.

Carbonyl compound	Ethynylation product	Conversion/%
$(CH_3)_2CO$	3-Methyl-1-butyn-3-ol	95
$C_2H_5COCH_3$	3-Methyl-1-pentyn-3-ol	67
$(CH_3)_2CHCH_2COCH_3$	3,5-Dimethyl-1-hexyn-3-ol	47
$C_6H_{10}O$	1-Ethynylcyclohexanol	71
$C_6H_5COCH_3$	3-Phenyl-1-butyn-3-ol	58
CH_3CHO	1-Butyn-3-ol	52
$(CH_3)_2CHCHO$	4-Methyl-1-pentyn-3-ol	72
C_3H_7CHO	1-Hexyn-3-ol	53
$C_4H_9CH(C_2H_5)CHO$	4-Ethyl-1-octyn-3-ol	75

The reaction mechanism of catalytic ethynylation in the presence of liquid ammonia–alkali metal hydroxide is as follows:

$$CH{\equiv}CH + NH_3 \longrightarrow HC{\equiv}C^- + NH_4^+$$

$$HC{\equiv}C^- + KOH \longrightarrow HC{\equiv}CK$$

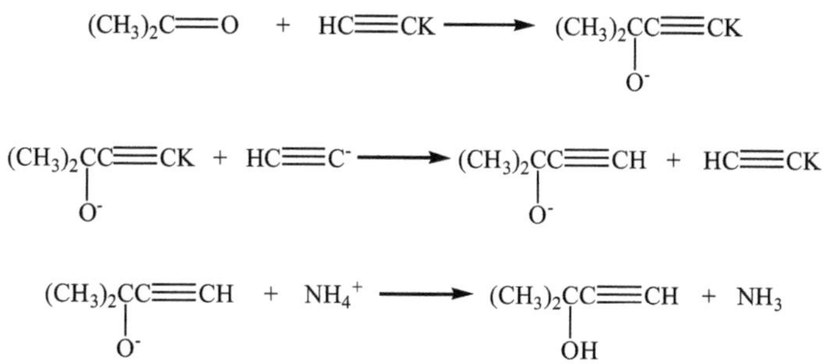

From the reaction mechanism, liquid ammonia is not only a solvent, but also an active substance. It participates in the reaction process, and forms acetylenic carbanion with acetylene and acts as a cocatalyst.

1.2.3.2 Continuous liquid-phase process for acetylene reactions

Gaseous acetylene exits the dangers of decomposition and explosion at high temperature and pressure, but the dissolved acetylene in a liquid phase is stable and safe at high temperature and pressure. Continuous liquid-phase reaction technology under high pressure changes the gas–liquid heterogeneous reaction into a homogeneous reaction of a single liquid phase and uses a tubular reactor for the reaction. In the reactor, acetylene is completely dissolved in the liquid phase, and there is no gas phase of acetylene present in the reactor. The reactants may be used to dissolve the acetylene, but if the solubility of the acetylene in the reactants is small, a solvent may also be employed. Before acetylene enters the reactor, it is dissolved in the liquid reactants (or reactants plus a solvent) at a temperature below room temperature under a pressure of 100–300 p.s.i.g. [19]. The solution together with catalyst solution is passed through a tubular reactor under high pressure. The reaction liquid from reactor outlet is cooled and decompressed to atmospheric pressure and subjected to fractional distillation for separation of the product and unreacted raw materials. Compared to gas–liquid reaction process, the liquid phase process is much safer and has a much higher space–time yield. However, dissolving acetylene in liquid phase under a pressure of 100–300 p.s.i.g. still requires an acetylene compressor, and the acetylene tail gas needs to be washed and recycled.

Li's research group in Xiamen University further improves the liquid-phase process [20]. Acetylene is dissolved in the liquid reactants (or reactants plus a solvent) at room temperature and atmospheric pressure in advance. The resulting solution is subjected to a liquid-phase cyclic reaction using a tubular reactor under high pressure. Figure 1.2 is a schematic diagram of the continuous liquid-phase cyclic reaction process for acetylene reaction under high pressure. The reaction device is provided with an absorber, a gas–liquid separator, a metering pump, a constant temperature oil bath, a tubular reactor, a cooler and a pressure relief valve. Acetylene is dissolved in a solution containing a catalyst and a reactant at room temperature and atmospheric pressure to form a saturated solution, and then the satu-

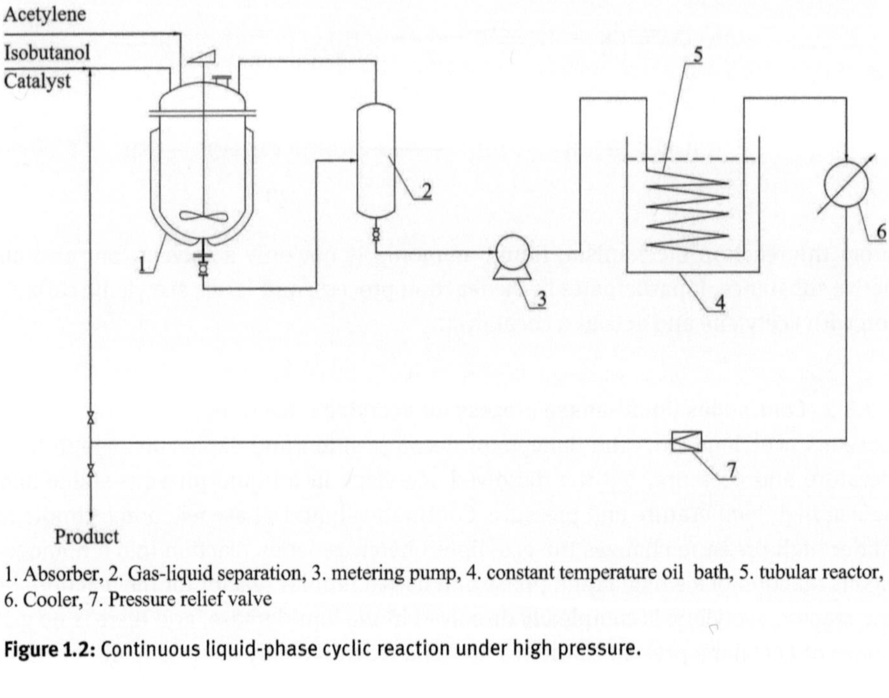

1. Absorber, 2. Gas-liquid separation, 3. metering pump, 4. constant temperature oil bath, 5. tubular reactor, 6. Cooler, 7. Pressure relief valve

Figure 1.2: Continuous liquid-phase cyclic reaction under high pressure.

rated solution is pressurized at 6–8 MPa by a metering pump and passed through a tubular reactor immersed in a constant temperature oil bath. The reaction liquid from reactor outlet is cooled and then decompressed to atmospheric pressure by a pressure relief valve. A small portion of the reaction liquid is discharged as the product and the rest of the liquid is circulated back to the absorber for absorption of acetylene together with the feed liquid under atmospheric pressure.

The main advantages of continuous liquid-phase cyclic reaction technology for acetylene reactions under high pressure are as follows:

1. In the reactor, the reaction system is a liquid phase, and there is no acetylene gas phase, thereby avoiding the danger of gas-phase acetylene decomposition and explosion under high temperature and pressure, and ensuring the safety of the production process.
2. The reaction system is a single liquid phase so that is not restricted by gas–liquid mass transfer, which increases the reaction rate and conversion.
3. The tubular reactor used is simple in structure, small in size, low in cost, easy to scale up to the industrial scale and avoids the amplification effect.
4. The heat transfer characteristics of the reactor are excellent, and the internal temperature of the reactor is uniform.
5. There is no need for an acetylene compressor, and the gas is not circulated so that the energy consumption is greatly reduced.
6. The reaction residence time is very short, and accordingly the amounts of by-products are greatly reduced.
7. No acetylene tail gas is generated in the crude product.

1.2.3.3 Gas–liquid reaction process for acetylene reactions

Gas–liquid reaction is a common reaction in chemical production. In general, for fast reactions, the gas–liquid reaction rate depends on the gas–liquid mass transfer rate, which can be represented by the following equation:

$$r_A = k_L a\left(C^* - C_L\right)$$

where r_A is the mass transfer rate, k_L is the mass transfer coefficient, a is the gas–liquid contact area, C^* is the equilibrium concentration of acetylene in the liquid phase under the system pressure and C_L is the concentration of acetylene in the liquid phase.

The mass transfer rate between gas and liquid is related to the properties of the system, gas–liquid contact area, gas solubility, equipment structure and other factors. For a specific system, the gas–liquid reaction rate can only be enhanced by selecting appropriate equipment type and process conditions for promotion of mass transfer. At present, three types of gas–liquid reactors including stirred tank reactor, bubble column reactor and loop reactor can be used for gas–liquid reaction. Their respective characteristics are shown in Table 1.7.

Gas–liquid reactions of acetylene are characterized by dangerous gas acetylene, exothermic reaction process and often need to operate at high temperature and pressure, and the reaction rate is controlled by mass transfer. Obviously, the use of loop reactor has obvious advantages in safety and efficiency, but it is also easy to scale up in the industry.

Table 1.7: Comparison of different gas–liquid reactors.

Reactor	Principle of enhanced mass transfer	Advantage	Disadvantage
Stirring kettle	Enhance the liquid turbulence for increase of mass transfer coefficient, k_L; large bubble breakage by stirring increases the specific surface area	Wide application range; fast mass transfer rate; good mixing effect	High energy consumption; hidden danger of leakage in the mechanical stirring kettle; expensive equipment for magnetic stirring kettle; batch operation; large deviation after scale-up
Bubble tower	Maintains high gas-phase host concentration and increases gas–liquid contact area for mass transfer by bubbling	Simple equipment, no internal components; easy to seal; energy saving; continuous operation	Poor mixing; uneven heat transfer; large deviation after scale-up
Loop reactor	Relies on the interaction of the circulation and the distributor, increases the mass transfer coefficient and the gas–liquid contact area	Simple equipment; easy to seal; circulation flow, good mixing, fast heat transfer; continuous operation; easy to enlarge the design	Internal components are required inside the reactor

References

[1] Tedeschi R J. Acetylene-based Chemicals from Coal and Other Natural Resources, Marcel Dekker, Inc, New York, USA, 1982.

[2] Schobert H. Production of acetylene and acetylene-based chemicals from coal, Chemical Reviews 2014, 114, 1743–141.

[3] Xiong M. Production of Calcium Carbide and its Deep-processed Products, Chemical Industry Press, Beijing, China, 2001.

[4] Huang X, Li S. Solubility of acetylene in alcohols and ketones, J. Chem. Eng. Data 2018, 63, 2127–2134.

[5] Pan Z, Lin Y, Wang H. Dissolved acetylene production process and safety, Liaoning Chemical Industry 2009, 38(09), 685–687.

[6] Zheng C. Safety measures related to acetylene, Medical Engineering Design 1985, 6, 26–32.

[7] Tang H, Mao Z. Safety measures for high-pressure acetylene in industrial reactor systems, Chemical Engineer 1991, 22(4), 48–31.
[8] Zou D. Discussion on the safe transportation of acetylene pipelines, Shihezi Technology 2012, 2, 31–33.
[9] Yang Z, Liu Y, Wang C. Improvement of the production process of acetylene produced by calcium carbide method, Polyvinyl chloride 2010, 02, 17–20.
[10] Shan T. Process analysis of safe production of dissolved acetylene, Low temperature and special gas 2010, 5, 44–48.
[11] Yan M. Hazardous factors analysis and safety control in the acetylene production process, Polyvinyl chloride 2005, 12, 40–43.
[12] Gao J. Acetylene production methods and technological progress, Natural Gas Chemical Industry 2005, 30(1), 63–66.
[13] Wang G, Yu X. Fire prevention measures for the production of acetylene by calcium carbide method, China Chlor-Alkali 2010, 4, 38–40.
[14] Song W, Zhang B, Zhou F, Ma D. The effect of acetylene properties on the safe storage and use of acetylene, Guangdong Chemical Industry 2014, 41(9), 115–116.
[15] Wang K, Xu Q. Analysis and prevention of acetylene safety water seal explosion accident, Chemical Production and Technology 2008, 15(2), 51–53.
[16] Cao H, Zhang J, Chen D. Safety and technological transformation in the production of acetylene, Polyvinyl chloride 2004, 5, 14–24.
[17] Ren H. Intrinsic safety design of acetylene pipeline transportation, Chemical Engineering Design 2014, 24(4), 42–45.
[18] Guo S. Using HAZOP analysis method to assess the safety risk of dry acetylene equipment, Polyvinyl Chloride 2015, 43(1), 42–44.
[19] Nedwick J J. Liquid phase process for acetylene reactions, I & E C Process Design and Development 1962, 1(2), 137–141.
[20] Li S F. An installation and method for preparation of isobutyl vinyl ether, CN, 201210570442.0, 2014.

Chapter 2
Derivatives from acetylene reacting with formaldehyde

2.1 1,4-Butanediol

1,4-Butanediol, abbreviated as BDO, also known as tetramethylene glycol, 1,4-butylene glycol, 1,4-dihydroxybutane, 1,4-tetramethylene glycol, butane-1,4-diol, butanediol, tetramethylene 1,4-diol and butylene glycol, is a derivative of acetylene and formaldehyde. CAS Registry Number: 110-63-4. Molecular formula: $C_4H_{10}O_2$, molecular weight: 90.1210. The chemical structure is

2.1.1 The properties of 1,4-butanediol

BDO is a colorless, oily and viscous liquid with a boiling point of 228 °C. It is very hygroscopic and miscible with water. It is soluble in methanol, ethanol and acetone, slightly soluble in ether, but almost insoluble in solvents such as aliphatic hydrocarbons, aromatic hydrocarbons and chlorinated hydrocarbons. BDO is a low-toxic chemical, but when it comes into contact with injured skin or if it is accidentally consumed, anesthesia will occur, which then causes sudden death due to paralysis of the central nervous system. The physical properties of BDO are shown in Table 2.1.

Table 2.1: Physical properties of 1,4-butanediol.

Melting point/°C	20.1	Flash point (open cup)/°C	121
Boiling point/°C		Heat of combustion/kJ · mol^{-1}	2,518.8
101.3 kPa	228	Enthalpy of vaporization/ kJ · mol^{-1}	
21.08 kPa	180	131.4 °C	68.2
13.3 kPa	171	193.2 °C	59.4
4.08 kPa	140	Enthalpy of fusion (20.4 °C)/ kJ · mol^{-1}	18.7
1.33 kPa	123	Heat capacity/kJ · kg · K^{-1}	
0.43 kPa	100	20 °C	2.2
0.133 kPa	86	50 °C	2.46
0.031 kPa	60	100 °C	2.9
Density/g · cm^{-3}		150 °C	3.33
20 °C	1.0171	Thermal conductivity (30 °C)/W · m^{-1} · K^{-1}	0.21

https://doi.org/10.1515/9783110714999-002

Table 2.1 (continued)

25 °C	1.0154	Viscosity (25 °C)/mPa · s	71.5
Refractive index (25 °C)	1.4445	Surface tension (22 °C)/N · m^{-1}	45.27 × 10^{-3}
Dielectric constant (20 °C)	31.1	Dipole moment/(×10^{-30} C · m)	2.40

BDO can be dehydrogenated in the presence of a catalyst such as copper-zinc-aluminum to produce γ-butyrolactone, which is one of the most important industrial derivatives of BDO. At high temperature, BDO is stripped off one water molecule to produce tetrahydrofuran in the presence of an acidic catalyst. As an alcohol, it is easy to be esterified with a carboxylic acid, for example, BDO and terephthalic acid are esterified to form dibutyl terephthalate, which can be used to manufacture engineering plastics. Under mild conditions, BDO reacts with isocyanate to form a polyurethane. Moreover, BDO can also undergo an oxidation reaction. Different oxidation methods obtain different products; for example, gas phase catalytic oxidation of BDO leads to maleic anhydride. However, liquid phase oxidation of BDO by air in an aqueous solution results in succinic acid.

2.1.2 Process for manufacture of 1,4-butanediol

At present, there are five industrialized routes for BDO production: (1) Reppe process, (2) butadiene/acetic acid process, (3) propylene oxide process, (4) direct hydrogenation of *n*-butane or maleic anhydride, (5) esterification and hydrogenation of *n*-butane or maleic anhydride. However, the mainstream processes are Reppe process and the direct hydrogenation of *n*-butane or maleic anhydride. Nowadays in the world, more than 40% of the total production of BDO is manufactured by Reppe process and more than 30% the total production of BDO is manufactured by the direct hydrogenation of maleic anhydride. In China, based on the current status of raw materials and production technology, modified Reppe process has been used for BDO production by most of the producers [1, 2].

2.1.2.1 Main chemical reactions

Reppe process, also known as acetylene-formaldehyde process, includes two steps: the first step is the reaction of acetylene with formaldehyde to form 1,4-butynediol and the second step is the catalytic hydrogenation of 1,4-butynediol to obtain BDO. The specific reaction process is as follows:

In the first step, acetylene reacts with formaldehyde:

$$HC{\equiv}CH + HCHO \longrightarrow HC{\equiv}CCH_2-OH$$

$$HC{\equiv}CCH_2-OH + HCHO \longrightarrow HO-CH_2C{\equiv}CCH_2-OH$$

Overall reaction:

$$HC{\equiv}CH + 2HCHO \longrightarrow HO-CH_2C{\equiv}CCH_2-OH$$

In the second step, 1,4-butynediol reacts with hydrogen:

$$HO-CH_2C{\equiv}CCH_2-OH + H_2 \longrightarrow HO-CH_2CH{=}CHCH_2-OH$$

$$HO-CH_2CH{=}CHCH_2-OH + H_2 \longrightarrow HO-CH_2CH_2CH_2CH_2-OH$$

Major side reaction:

$$HC{\equiv}CCH_2-OH + 2H_2 \longrightarrow CH_3CH_2CH_2-OH$$

$$n HC{\equiv}CH \longrightarrow \text{Various acetylene oligomers}$$

$$OH-CH_2CH{=}CHCH_2-OH \xrightarrow{\text{Isomerization}} HO-CH_2CH_2CH_2CHO$$

Reppe process is the earliest production method developed for BDO. This process has been improved. It is very suitable for large-scale production of BDO. The advantages include mature technology, low production cost, short process, high product yield, less by-products and low investment. Therefore, it is still the mainstream process for production of BDO in the world.

2.1.2.2 Reaction of acetylene with formaldehyde to 1,4-butynediol

In the initial Reppe process, a gas–liquid–solid three-phase trickle-bed reactor was used. Acetylene and aqueous formaldehyde solution flowed down in parallel through the cuprous acetylide catalyst bed. The particle size of the catalyst was in the range from 3 to 7 mm. The reaction was carried out at a temperature between 90 and 110 °C. In general, the partial pressure of acetylene was controlled to be from 0.5 to 0.6 MPa. The process had a large processing capacity with low operating cost. Furthermore, there was no problem for separation of the catalyst and the product. However, it tended to form localized hot spots in the trickle-bed reactor, which degraded catalyst performance. High pressure of acetylene in the reaction system results in a risk of explosion. For this reason, it was required a high safety factor in the design of the processing equipment and pipelines, and even the filling of the equipment and pipelines with stainless steel packings or stainless steel tubes with small diameter to ensure safety. The catalyst used in the reaction was cuprous acetylide, which was very unstable and prone to explosion. The process has now been replaced by a series of modified Reppe processes that employ a slurry bed or a suspended bed reactor. Figure 2.1 shows a process flow diagram for 1,4-butynediol by the modified Reppe process.

The slurry bed technology was developed by ISP Company in the United States. Different from the gas–liquid–solid mixing system in the trickle bed, aqueous formaldehyde solution and catalyst particles were mixed in the slurry bed reactor to form a slurry that was used as a continuous phase, and then a discontinuous phase of acetylene entered the slurry system from the bottom of the liquid layer. Mechanical stirring, bubble stirring or ultrasonic dispersion was applied to mix the aqueous formaldehyde solution and the catalyst uniformly. When acetylene entered the slurry system from the bottom of the reactor, it was well dispersed in the slurry system by a sieve plate or a gas dispersion device. The heat of reaction was removed by means of an internal cooling coil in the reactor or an external circulation of the reaction mixture. The internal temperature distribution in the reactor was uniform and no local hot spot occurred; therefore, the operation was safe and reliable. The particle size of the catalyst in the slurry bed reactor was in the range from 0.05 to 0.2 mm. The pressure of acetylene was reduced to below 0.2 MPa, and the reaction temperature could be lowered to 80–90 °C. However, it was required to recover the catalyst from the reaction mixture for recycling, resulting in an increase in the operation cost.

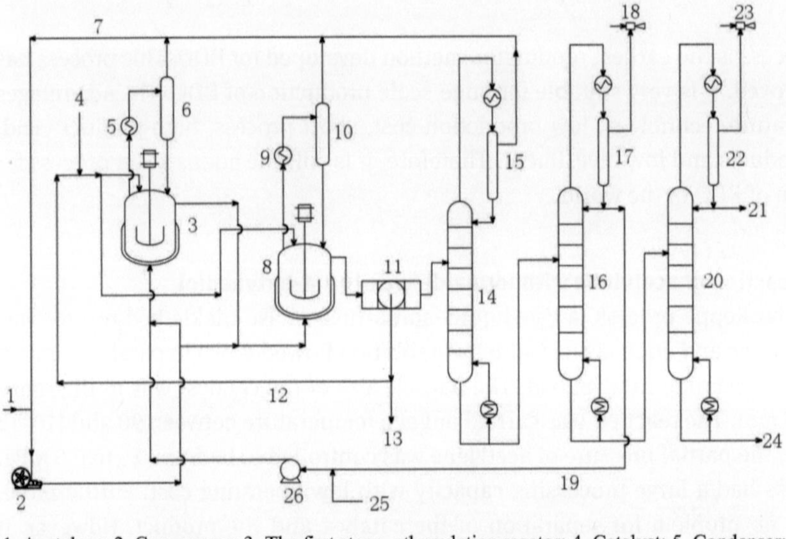

1. Acetylene; 2. Compressor; 3. The first stage ethynylation reactor; 4. Catalyst; 5. Condenser; 6, 10, 15, 17, 22. Gas-liquid separator; 7. Recycle acetylene; 8. The second stage ethynylation reactor; 9. Compressor; 11. Catalyst filter; 12. Recycle catalyst; 13. Spent catalyst; 14. Stripping tower; 16. Formaldehyde removal tower; 18. Steam jet pump; 19. Recycle formaldehyde; 20. Heavy component removal tower; 21. Butynediol; 23. Steam jet pump; 24. Heavy components; 25. Formaldehyde; 26. Pump

Figure 2.1: Process flow diagram for 1,4-butynediol by modified Reppe process.

To be different from ISP, BASF in Germany and DuPont in the United States adopted a suspension bed reaction technology, in which the catalyst particles were suspended by the flow of acetylene and formaldehyde solution. The catalyst did not flow out of the reactor with the reaction liquid. The pressure of acetylene was reduced to less than 0.2 MPa, and the reaction temperature could be lowered to 80–90 °C. The difference between the suspended bed and the slurry bed was that the catalyst was separated from the product in different manners. Catalyst separation was accomplished inside the reactor for the suspended bed, while catalyst separation was accomplished outside the reactor for the slurry bed. No catalyst agglomeration and deactivation occurred in the suspended bed, so that this technology was more stable and reliable, and the operation was simple.

The main active ingredient of the catalyst used for the reaction of acetylene and formaldehyde was copper, and bismuth was used as a cocatalyst. The Cu–Bi catalyst was loaded on a support. Commonly used supports were SiO_2, SiO_2–MgO, Al_2O_3, molecular sieves, spinel and activated carbon. The preparation of the catalyst was usually carried out by solution impregnation or coprecipitation of a water-soluble salt of the active component. The copper content was about 20% and the bismuth content is 2–3% in the catalyst. As the copper content increased from 13% to 26%, the catalyst activity was multiplied. The catalyst required to be activated before use. The activation was carried out in situ in the reactor by introducing a dilute solution of formaldehyde and an acetylene of 0.1–0.15 MPa at 60–100 °C for about 10 h to convert the copper ion into cuprous acetylide. During the ethynylation reaction, the cuprous acetylide might be reduced to copper of zero valence, which promoted the polymerization of acetylene, resulting in the formation of a by-product polyacetylene. The resultant polyacetylene adhered to the catalyst surface, causing catalyst poisoning and inactivation. Therefore, the catalyst life was reduced. The addition of bismuth could effectively prevent the copper (I) from reduction to copper of zero valence.

2.1.2.3 Hydrogenation of 1,4-butynediol

After ethynylation, the product was subjected to fractional distillation to remove the light components and the heavy components, then treated with an ion exchange resin to remove metal ions and colloids, and subsequently introduced into the hydrogenation reactor. A two-stage hydrogenation process was used by ISP for hydrogenation of 1,4-butynediol to BDO. The first stage of hydrogenation was carried out in a stirred tank reactor with a Raney–Ni catalyst at low temperature and low pressure. Aqueous solution of 1,4-butynediol with a concentration of 35–40% was introduced into the reactor at 50–60 °C under a hydrogen pressure of 1.4–2.1 MPa. After the first stage of hydrogenation, the reaction mixture was discharged from the reactor and separated into gas and liquid phases. Hydrogen was recycled and the liquid was sent to a filter to recover the catalyst. The filtrate was introduced into the second stage hydrogenation reactor for further hydrogenation. The second stage reactor was a gas–liquid–solid three-phase trickle-bed reactor filled with a Ni-Cu–Mn/silica gel catalyst. The reaction temperature was 120–140 °C and the hydrogen pressure was 14–21 MPa. After reaction and gas–liquid separation, hydrogen was recycled and the liquid was subjected to fractional distillation to obtain BDO of a purity more than 99.5%. The process flow diagram for hydrogenation of 1,4-butynediol to BDO is displayed in Figure 2.2. The yield of BDO is greater than 90%. As the support for catalyst was changed from silica gel to γ-Al_2O_3, the catalyst activity was improved and the yield of BDO was increased from 90% to 95%.

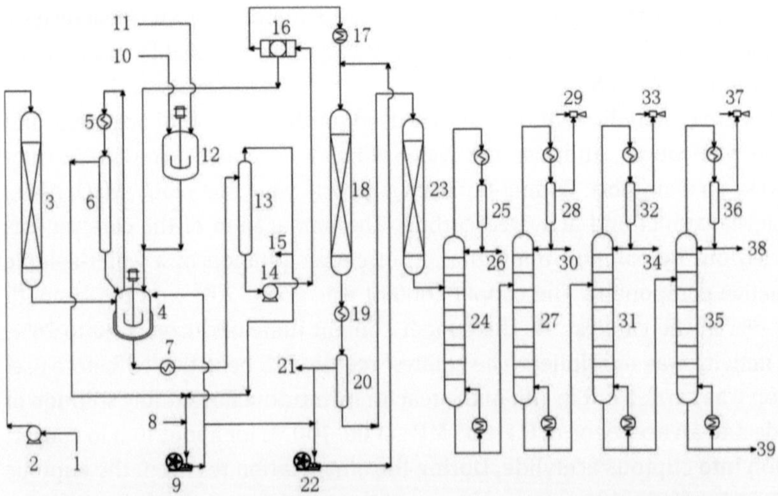

1. 1, 4–Butynediol; 2, 14. Pump; 3, 23. Purification tank; 4. The first stage hydrogenation reactor; 5. Condenser; 6, 13, 20. Gas-liquid separator; 7, 19.Cooler; 8. Hydrogen; 9, 22. Compressor; 10. Raney Ni catalyst, NaOH and water; 11. Nitrogen; 12. Catalyst activation tank; 15, 21. Vent to burn; 16. Filter; 17. Preheater; 18. The second stage hydrogenation reactor; 24, 31. Light components removal tower; 25, 28, 32, 36. Reflux tank; 26, 34. Light components; 27. Water removal tower; 29, 33, 37. Steam jet pump; 30. Water; 35. Heavy components removal tower; 38. 1,4-butanediol; 39. Heavy components.

Figure 2.2: Process flow diagram for hydrogenation of 1,4-butynediol to BDO.

2.1.3 Uses and consumption of BDO

BDO is mainly used to manufacture tetrahydrofuran, γ-butyrolactone, engineering plastic polybutylene terephthalate (PBT), polyurethane elastomers, biodegradable plastic polybutylene succinate and hydroxybutyl vinyl ether. The largest consumption of BDO in the world is for the production of tetrahydrofuran. In addition, BDO is also used as a solvent and pharmaceutical intermediate.

In recent years, the consumption of BDO in the world has increased steadily. In 2013, the consumption was 1.62 million tons, and in 2017 it increased to 2.102 million tons. Among them, about 51.9% of BDO was used for the production of tetrahydrofuran, 22.5% of the BDO was used for PBT resin for household appliance and color TV industry, 12.6% of the BDO was for the production of polyurethane elastomers, 11.4% of the BDO was for the production of γ-butyrolactone and 1.6% of the BDO was for other uses.

Before 2006, the production output of BDO in China was less than the import volume, indicating the market demand main reliance on imports. The self-sufficiency percentage of BDO exceeded 50% for the first time in 2006 and reached 97% in 2015, as listed in Table 2.2. China is currently the world's largest consumer of BDO. The consumption of BDO in the mainland of China was 973,000 tons, accounting for 46.29% of the world's total consumption in 2017.

Table 2.2: Consumption and self-sufficiency of BDO in China during the years from 2004 to 2017.

Year	Consumption (10,000 t a^{-1})	Self-sufficiency (%)
2004	13	34
2005	14.6	42
2006	17	54.7
2007	23.4	54.7
2008	26	59
2009	32	72
2010	36	76
2011	40	80
2012	48	86.7
2013	52	92
2014	61	95
2015	72	97
2016	82	97.6
2017	97.3	97.6

2.1.4 Production status of 1,4-butanediol

In recent years, global BDO production capacity has rapidly increased, which is stimulated by the developments of the downstream products such as tetrahydrofuran, γ-butyrolactone, engineering plastic PBT and polyurethane elastomers. The global production capacity of BDO was 3.173 million tons in 2014, and it increased to 3.622 million tons in 2017. The production and sales are basically balanced. BASF is the largest producer of BDO in the world, with a total capacity of nearly 500,000 tons, and its production facilities are located in Europe, America and Asia.

The production of BDO in China started late. Since China did not have the technology of large-scale industrialization of BDO production, almost all of the BDO required in China were imported before the 1990s. In the 1990s, Shandong Shengli Oilfield introduced a 10,000 t a^{-1} of butane-based maleic anhydride plant from Italy, and 10,000 t a^{-1} of BDO installation by esterification and hydrogenation of maleic anhydride from the Davy Company in the United Kingdom. The BDO installation, which was successfully put into operation in 1996, ended the history of the inability to produce BDO on a large scale in China. A few years later, Shanxi Sanwei and other companies introduced Reppe process for the production of BDO. Since then, large companies outside of China were optimistic about the Chinese market, and begun to transfer technology to China or invest in the construction of BDO production facilities. In the past 10 years, the production of BDO was developing rapidly, and its production capacity and market demand were increasing rapidly year by year. Currently, the product output of BDO has leapt to the first place in the world. Figure 2.3 represents the growth of BDO production capacity in China from 2004 to 2017. The BDO production capacity in China was up to 1.94 million tons in 2017, which accounted for 53.6% of global production [3].

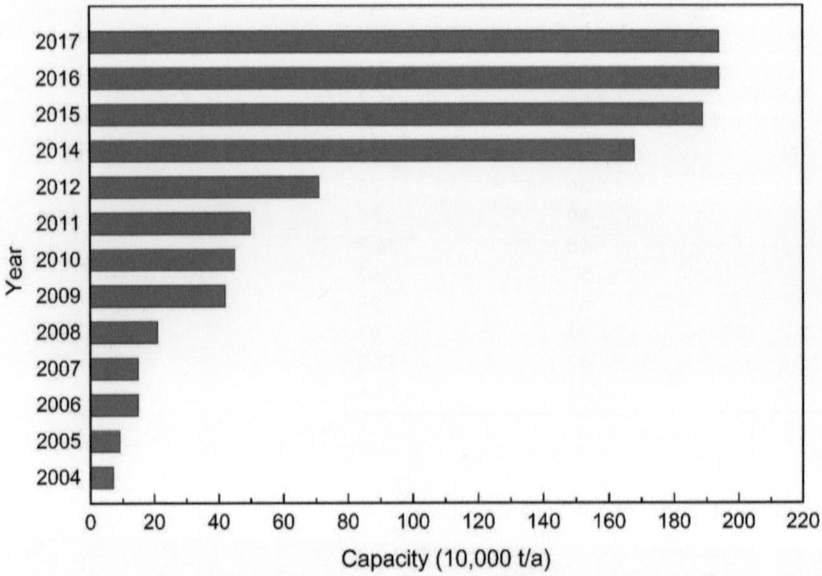

Figure 2.3: Production capacity of BDO in China during the years from 2004 to 2017.

There are about 20 manufacturers of BDO in China. The production technology is mainly Reppe process. Table 2.3 lists the main BDO manufacturers and each production capacity by Reppe process in 2017. Statistics showed that the production capacity of BDO by Reppe process was 1.71 million tons, which accounted for about 88% of the total capacity in China. The process by esterification and hydrogenation of maleic anhydride was used by some manufacturers, but it is now eliminated due to high production cost and low economic efficiency. The propylene oxide process is adopted by Dalian Chemical (Jiangsu) Co., Ltd. (production capacity of 80,000 t a^{-1}) and Changchun Chemical (Panjin) Co., Ltd. (production capacity of 150,000 t a^{-1}).

The Reppe process is currently the mainstream process for the production of BDO in China, and its cost depends to a large extent on the source and price of the

Table 2.3: China manufacturers and production capacity of BDO by Reppe process in 2017.

No	Manufacturers	Capacity, 10,000 t /a
1	Xinjiang Meike Chemical Co., Ltd.	26.0
2	Xinjiang Tianye (Group) Co., Ltd.	21.0
3	Xinjiang Guotai Xinhua Mining Co., Ltd.	20.0
4	GD Sinopec Ningxia Energy Chemical Industry Co., Ltd.	20.0
5	Shaanxi Bidiou Chemical Co., Ltd.	13.0
6	Henan Coal Chemical Group Hemei Company	10.0
7	Inner Mongolia Dongyuan Technology Co., Ltd.	10.0
8	Xinjiang Lanshan Tunhe Energy Co., Ltd.	10.0
9	Henan Kaixiang Fine Chemical Co., Ltd.	9.0
10	Sichuan Tianhua Co., Ltd.	8.5
11	Shanxi Sanwei Group Co., Ltd.	7.5
12	Chongqing Jianfeng Chiyuan Chemical Co., Ltd.	6.0
13	Shaanxi Shenmu Guorong Fine Chemical Co., Ltd.	6.0
14	Fujian Meizhou Bay Chlor-Alkali Industry Co., Ltd.	4.0
	Total	171.0

starting material acetylene. When the oil price is low, the *n*-butane/maleic anhydride method using *n*-butane as a raw material is more competitive. When the oil price is above 60 US dollars/barrel, due to the relationship between *n*-butane and oil price, the *n*-butane/maleic anhydride method is not as economic as the acetylene method.

Comparison of raw material cost in different BDO production routes is summarized in Table 2.4.

Table 2.4: Comparison of raw material cost in different BDO production routes.

Process	Raw material	Consumption	Price/CNY	Cost/CNY
Reppe process	Calcium carbide	1.2	3,400	4,080
	Formaldehyde	2	1,300	2,600
	Hydrogen	0.2	10,000	2,000
	Total			8,680
Maleic anhydride	Maleic anhydride	1.1	9,700	10,670
	Methanol	0.7	2,500	1,750
	Hydrogen	0.2	10,000	2,000
	Total			14,420

The above analysis shows that the price changes of natural gas, calcium carbide and ethylene are different. The price of ethylene is greatly affected by the international crude oil market. According to the cost of oil production of Sinopec 60 US dollars barrel^{-1}, the corresponding ethylene price is 7,320 yuan t^{-1}. The price of calcium carbide at 2,700 yuan t^{-1} is equivalent to 1.4 yuan m^{-3} of nature gas. As long as the crude oil is more than 60 US dollars barrel^{-1}, the advantages of the acetylene method will be significant. In terms of natural gas prices, China recently introduced a natural gas price reform plan to gradually increase the price of natural gas and rationalize the price relationship with alternative energy sources. At present, based on the equivalent calorific value, the price ratio of oil: natural gas: coal = 1:0.24:0.17 in China, while oil: natural gas: coal = 1:0.6:0.2 in the international market, therefore, with the existing calcium carbide acetylene resources to develop BDO by acetylene route has lower cost and stronger competitiveness.

2.2 Tetrahydrofuran

Tetrahydrofuran, also known as oxacyclopentane, oxolane, tetramethylene oxide, 1,4-epoxybutane, is a derivative of BDO. CAS registry number: 109-99-9. Molecular formula: C_4H_8O, molecular weight: 72.1057. The chemical structure is

2.2.1 Properties of tetrahydrofuran

Tetrahydrofuran is a colorless and transparent liquid with an ether-like odor. It is a volatile liquid having a boiling point of 66 °C under atmospheric pressure. It is miscible with water and common organic solvents such as alcohol, ether, ester, ketone, aliphatic hydrocarbon, aromatic hydrocarbon and halogenated hydrocarbon. Many polymer materials can be dissolved in tetrahydrofuran. It forms an azeotrope with water, having an azeotropic point of 64 °C under atmospheric pressure and a water content of 5.6% (mass fraction) in the azeotrope. As the pressure is increased to 0.8 MPa, the azeotropic point is changed to 106 °C, and the water content in the azeotrope is increased to about 9%. Using this property in industry, anhydrous product of tetrahydrofuran can be obtained by changing the pressure in the distillations to get rid of its moisture content. The physical properties of tetrahydrofuran are summarized in Table 2.5.

Table 2.5: Physical properties of tetrahydrofuran.

Melting point/°C	−108.5	Enthalpy of vaporization/$kJ \cdot kg^{-1}$	413
Boiling point/°C		Enthalpy of fusion/$kJ \cdot mol^{-1}$	8.54
15.20 kPa	15	Constant pressure heat capacity/$kJ \cdot kg^{-1} \cdot K^{-1}$	
23.46 kPa	25	$C_{p,liquid}$ (20 °C)	1.96
35.07 kPa	35	$C_{p,liquid}$ (25 °C)	1.72

Table 2.5 (continued)

51.33 kPa	45	$C_{p,\text{liquid}}$ (60 °C)	1.55
73.33 kPa	55	$C_{p,\text{gas}}$ (76 °C)	1.26
101.3 kPa	66	$C_{p,\text{gas}}$ (126 °C)	1.47
Density (20 °C)/g · cm^{-3}	0.8892	Dielectric constant (25 °C)	7.58
Refractive index (25 °C)	1.4040	Surface tension (25 °C)/N · m^{-1}	26.4 × 10^{-3}
Viscosity (20 °C)/mPa · s	0.55	Dipole moment/(×10^{-30} C · m)	5.70
Flash point (open cup)/°C	17.2	Critical temperature/°C	26.8
Heat of combustion/kJ · g^{-1}	34.74	Critical pressure/MPa	5.19

Tetrahydrofuran is a first-grade flammable product. It can be mixed with air to produce an explosive mixture with an explosion limit of 1.8% to 11.8%. When tetrahydrofuran is in contact with air, it is easy to oxidize with oxygen in the air to form peroxide. Therefore, air should be isolated during the process of production, storage, transportation and processing, and the sealed container should be filled with nitrogen gas. The antioxidant 2,6-di-*tert*-butyl-4-methylphenol of 100–200 mg kg^{-1} is often added to tetrahydrofuran as a stabilizer. When tetrahydrofuran is oxidized with nitric acid, succinic acid is formed. Under the catalysis of alumina, tetrahydrofuran is reacted with ammonia at a temperature of 300 to 400 °C to obtain pyrrolidine. Tetrahydrofuran reacts with hydrogen sulfide to form tetrahydrothiophene. In the presence of zinc chloride, it reacts with an acid or an acid chloride to easily open the ring and form BDO or 1,4-dihalide. Chlorination of tetrahydrofuran at room temperature under the light results in the formation of 2,3-dichlorotetrahydrofuran. Using acid phosphate as a catalyst, tetrahydrofuran is dehydrated at 270 °C to form butadiene. Tetrahydrofuran is hydrolyzed and ring-opened to form a polyether diol in the presence of a strongly acidic catalyst.

2.2.2 Process for manufacture of tetrahydrofuran

At present, the production methods of tetrahydrofuran mainly include furfural route, BDO route, butadiene oxidation and catalytic hydrogenation of maleic anhydride. Among them, the BDO route is adopted by most manufacturers in the world for the production of tetrahydrofuran.

The BDO route, also known as catalytic dehydration and cyclization of BDO, or Reppe process, was successfully developed by German IG company and Dr. Reppe. Acetylene and formaldehyde are used as starting materials to produce BDO, and then BDO is dehydrated in the presence of an acidic catalyst to obtain tetrahydrofuran:

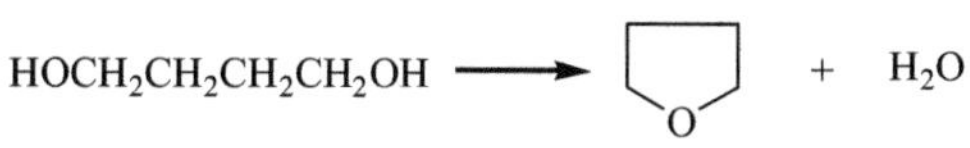

The acidic catalyst used includes sulfuric acid, phosphoric acid, heteropoly acid, alumina, zirconia, molecular sieves such as ZSM-5, and strong acid ion exchange resins, Al^{3+}-treated montmorillonite and the like. Sulfuric acid is the earliest catalyst used in the industrial production of tetrahydrofuran, and it is also a widely used catalyst. The advantages of sulfuric acid catalytic process include mature technology, relatively simple process, low reaction temperature and high yield of tetrahydrofuran. However, sulfuric acid is easy to corrode equipment, and the resulting wastewater pollutes the environment. Heteropolyacids and montmorillonite catalysts are still in the research stage and are not used for industrial production of tetrahydrofuran. The alumina solid catalyst has the advantages of low cost, no corrosion and high selectivity in the dehydration reaction, but a high reaction temperature up to 320 °C is required, which increases equipment investment and energy consumption. So far, no reports of industrialization of alumina catalyst for the production of tetrahydrofuran have been reported. Currently, sulfuric acid and strong acid ion exchange resin are the most popular catalysts for the industrial dehydration of BDO to produce tetrahydrofuran [4–6].

The dehydration reaction can be carried out in the gas phase or in the liquid phase, either intermittently or continuously. The starting material BDO used may be pure BDO or a fraction rich in BDO in other processes. For example, in the production process of BDO by Reppe process, after hydrogenation of butynediol and removal of water from the reaction solution, the crude BDO can directly be used as the raw material for production of tetrahydrofuran. High-purity product can also be obtained without purification of the crude BDO. In this way, the manufacture process can be simplified.

Usually, the dehydration reaction of BDO was carried out at 150 °C in a fixed-bed reactor filled with strong acid ion exchange resin. Available resin types included DOWEX-50, Amberlite IR-120 and Amberlyst 15. All of them were sulfonated styrene-divinylbenzene resins. Both the conversion of BDO and the selectivity to tetrahydrofuran were more than 99%.

The products of the dehydration reaction of BDO are mainly tetrahydrofuran, water, light and heavy organic by-products. The light and heavy components are easily removed by fractional distillation. Tetrahydrofuran forms an azeotrope with water, which makes the separation process difficult. There are two methods for removal of the water from aqueous tetrahydrofuran solution. One is extractive distillation using solvent such as ethylene glycol, propylene glycol and BDO. Another is pressure-sensitive distillation, which utilizes the feature that the azeotropic point and water content in the azeotrope increase with the elevation of pressure. Two fractional distillation towers are applied, in which one is operated under atmospheric pressure to obtain an azeotrope, and another is carried out under increased pressure to remove the water in the azeotrope, thereby a tetrahydrofuran product having a water content of less than 0.5% is obtained.

2.2.3 Uses of tetrahydrofuran

The main use of tetrahydrofuran is ring-opening polymerization to produce polytetrahydrofuran, which is used as a raw material for the production of polyurethane elastomers, polyurethane elastic fibers and polyurethane adhesives. Compared with other elastic materials, the elastic material prepared from polytetrahydrofuran has excellent stability against hydrolysis, gas permeability and wear resistance, and also exhibits good elasticity, flexibility and impact resistance at low temperatures. It has unique and broad application prospects in textiles, pipes, chemicals, medical equipment and so on. Currently, more than 50% of the tetrahydrofuran produced in the world is used for manufacture of polytetrahydrofuran.

Tetrahydrofuran is an organic solvent with excellent performance. It has a fast dissolution rate, good penetration and diffusion properties on the surface and inside of the resin; therefore, it has been widely used as a solvent for resins. It can dissolve all resins except polyethylene, polypropylene and fluororesin, especially for polyvinyl chloride, polyvinylidene chloride and its copolymers, and results in low-viscosity solution, so it can be used for polyvinyl chloride pipe processing, manufacture of surface coatings, adhesives and films as well as inks, extractants and artificial leather surface treatments. Tetrahydrofuran is also used for the cleaning of polymerization reactors for polyvinyl chloride production. Moreover, it is widely used as a solvent in Grignard reaction, polymerization reaction and $LiAlH_4$ reduction reaction.

Tetrahydrofuran is also used as an intermediate in organic synthesis and can be applied to synthesize many important fine chemicals such as tetrahydrothiophene, tetrahydrothiophenol, 1,4-dichlorobutane and pyrrolidine. In the pharmaceutical field, it is used as a raw material for the production of pentoxyverine.

2.2.4 Production status of tetrahydrofuran

The global production capacity of tetrahydrofuran is more than 1 million t a^{-1}, of which the United States and the European Union account for approximately 25% each and Asia accounts for approximately 50%. Many manufacturers of BDO also produce tetrahydrofuran and directly use tetrahydrofuran to produce downstream products such as polytetrahydrofuran, rather than directly sell tetrahydrofuran product to the market. Therefore, it is difficult to accurately calculate the global production capacity and actual output of tetrahydrofuran. In recent years, the production of tetrahydrofuran in the United States and the European Union has grown slowly or even shrunk, while the production capacity in Asia has grown rapidly. Asia is not only the main producer of tetrahydrofuran, but also the largest consumer in the world.

The large-scale industrial production of tetrahydrofuran in China began in the late 1990s. In 1998, Qiqihar Advance Chemical Factory adopted China's own technology of BDO dehydration to build a production installation with an annual output of 3,000 tons of tetrahydrofuran and then expanded to 7,000 t a^{-1}. In the same year, another production facility was also built in Shijiazhuang city, Hebei province, with an annual output of 1,000 tons of tetrahydrofuran. The raw material BDO used at that time was completely dependent on imports. Subsequently, Shandong Shengli Oilfield built a production unit of BDO from maleic anhydride, which produced 1,000 t a^{-1} of tetrahydrofuran as a by-product. Shanxi Sanwei Group started production in 2002 with the construction of a 15,000 t a^{-1} tetrahydrofuran unit by importing the technology from DPT Company (United Kingdom). In the past 20 years, the production of tetrahydrofuran was developed rapidly from scratch. In 2011, more than 10 manufacturers in China were engaged in the production of tetrahydrofuran and the total production capacity reached 294,000 tons, as shown in Table 2.6.

Table 2.6: China tetrahydrofuran production enterprises and capacity in 2011.

No.	Manufacturers	Capacity (1,000 t a^{-1})	Process
1	Qiqihar Qianjin Chemical Factory	7	1,4-Butanediol dehydration
2	Hebei Shijiazhuang Xinyu Sanyang Industrial Co., Ltd.	1	1,4-Butanediol dehydration
3	Shandong Dongying Jiatai Petrochemical Co., Ltd.	1	Davy 1,4-butanediol by-product
4	Shanxi Sanwei Group Co., Ltd.	15	1,4-Butanediol dehydration
5	Zibo Linzi Bingqing Fine Chemical Factory	1.4	1,4-Butanediol dehydration
6	Jiangsu Dalian Chemical	40	1,4-Butanediol dehydration
7	Shanghai Caojing BASF Chemical Company	80	Maleic anhydride method
8	Nanjing Bluestar Chemical New Material Co., Ltd.	20	Maleic anhydride method
9	Sinochem International Taicang Xingguo Industrial Co., Ltd.	20	1,4-Butanediol dehydration
10	China National Petroleum Jilin Qianguo Refinery	20	Furfural
11	Ningbo Mitsubishi Chemical Co., Ltd.	25	1,4-Butanediol dehydration
12	Sichuan Tianhua Co., Ltd.	50	1,4-Butanediol dehydration
13	Sinopec Yizheng Chemical Fiber Co., Ltd.	10	PBT by-product
	Total	294	

In recent years, as some large BDO plants have been put into operation one after another, the production of tetrahydrofuran has also made great progress in China. For example, companies including Xinjiang Meike Chemical Co., Ltd. and Chongqing Jianfeng Chiyuan Chemical Co., Ltd. built tetrahydrofuran and polytetrahydrofuran units. The total production capacity of tetrahydrofuran in China was about 379,000 tons in 2017.

2.3 γ-Butyrolactone

γ-Butyrolactone, abbreviated as GBL, also known as γ-hydroxybutyric acid cyclic ester and γ-hydroxybutyric acid lactone, is a derivative from dehydrogenation of BDO. CAS registry number: 96-48-0. Molecular formula: $C_4H_6O_2$, molecular weight: 86.0892. The chemical structure is

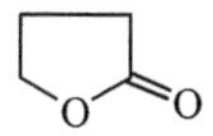

2.3.1 Properties of γ-butyrolactone

γ-Butyrolactone is a colorless, transparent and oily liquid with a certain hygroscopicity and a slight odor similar to acetone. Its density is 1.125 g cm^{-3} at 25 °C. The boiling point is 206 °C under atmospheric pressure. It is miscible with water, alcohols, ketones, esters, ethers and aromatics. In addition, it is an excellent solvent for many polymers. However, γ-butyrolactone is only slightly soluble in linear alkanes and cycloparaffins. The physical properties of γ-butyrolactone [7, 8] are shown in Table 2.7.

Table 2.7: Physical properties of γ-butyrolactone.

Melting point/°C	−45	Flash point (open cup)/°C	98.3
Boiling point/°C		Heat of combustion/kJ · mol^{-1}	2,010.6
0.133 kPa	35	Enthalpy of vaporization/ kJ · mol^{-1}	
1.33 kPa	77	84 °C	51.8
1.60 kPa	89	119 °C	49.5
13.3 kPa	134	134 °C	48.2
101.3 kPa	204	Enthalpy of fusion/kJ · mol^{-1}	9.57
Density (25 °C)/g · cm^{-3}	1.1254	Heat capacity/kJ · kg^{-1} · K^{-1}	
Refractive index (25 °C)	1.4348	20 °C	1.60
Viscosity (25 °C)/mPa · s	1.75	25 °C	1.68
Dielectric constant (25 °C)	39.10	100 °C	1.85
Surface tension (25 °C)/N · m^{-1}	44.6 × 10^{-3}	Critical temperature/°C	436
Dipole moment/(×10^{-30} C · m)	13.74	Critical pressure/MPa	3.4

γ-Butyrolactone is chemically active and can react with a variety of compounds, especially ring-opening reactions, or reactions where the oxygen atom on the ring is replaced by other heteroatoms.

2.3.2 Process for manufacture of γ-butyrolactone

At present, the main production processes of γ-butyrolactone include dehydrogenation of BDO (i.e., Reppe process using acetylene and formaldehyde as starting materials) and hydrogenation of maleic anhydride. Up to now, about 75% of the manufacturers produce γ-butyrolactone by dehydrogenation of BDO in vapor phase, and the next is hydrogenation of maleic anhydride to produce γ-butyrolactone. γ-Butyrolactone is a fine chemical intermediate with high added value, and the optimization of its production technology will bring obvious economic benefits.

2.3.2.1 Dehydrogenation of 1, 4-butanediol to γ-butyrolactone

The production of γ-butyrolactone by dehydrogenation of BDO was first developed and industrialized by BASF in Germany in 1946. So far, this method has been improved several times and is still the main method for industrial production of γ-butyrolactone. The scale of a single production unit has reached an annual output of 10,000 tons.

Dehydrogenation of BDO can be carried out in the liquid phase or in the gas phase, but in industry, the gas phase dehydrogenation is generally used. In the presence of a catalyst, BDO is first dehydrogenated to form γ-hydroxy butyraldehyde, and then, the latter is cyclized and dehydrogenated to form γ-butyrolactone. The reaction equation is as follows:

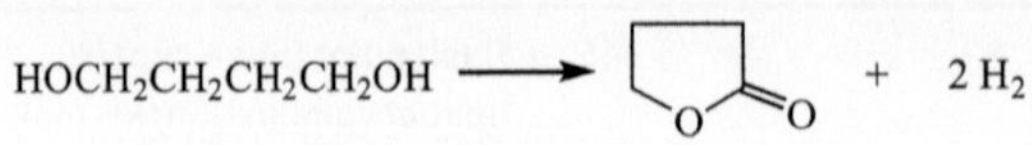

The catalyst used is a copper catalyst, the promoter is zinc, manganese, chromium and so on, and the catalyst carrier is alumina or silica gel. A typical catalyst composition is $CuO:ZnO:Al_2O_3 = 3:2:1$. In general, the catalyst is prepared by coprecipitation. A fixed-bed reactor filled with the catalyst is applied in industrial production, and the reaction is carried out at a temperature between 190 and 240 °C under atmospheric pressure [9].

The process flow diagram is shown in Figure 2.4. Fresh BDO and recycled BDO are heated by a preheater and sent to an evaporator for vaporization. The vapor of BDO is mixed with hydrogen, and the resulting gas mixture is introduced into the dehydrogenation reactor from the top. The reaction mixture passes through the catalyst bed, flows out from the bottom of the reactor, exchanges heat with the circulating hydrogen, cooled by a condenser and then subjected to a gas–liquid separator. The separated hydrogen is partially recycled, and the others are used as by-products. The separated liquid is sent to a light fraction removal tower, and by-products lighter than γ-butyrolactone are distilled overhead, which are mainly tetrahydrofuran, butanol, butyric acid, water and the like. These by-products enter the tetrahydrofuran fractional distillation tower, and an azeotrope of tetrahydrofuran and water is distilled off from the top, and other by-products are discharged from the bottom. The bottom product of the light fraction removal tower is mainly γ-butyrolactone and unreacted

BDO, which is sent to the γ-butyrolactone tower for fractional distillation under vacuum. γ-Butyrolactone product with a purity of more than 99% is obtained from the top. The bottom product of the γ-butyrolactone tower is BDO which is recycled. Both the conversion of BDO and the selectivity to γ-butyrolactone are more than 98%. The entire process is substantially free from environment pollution. The production unit can be made of carbon steel, but the part in contact with the hot reaction product containing butyric acid needs to be made of stainless steel.

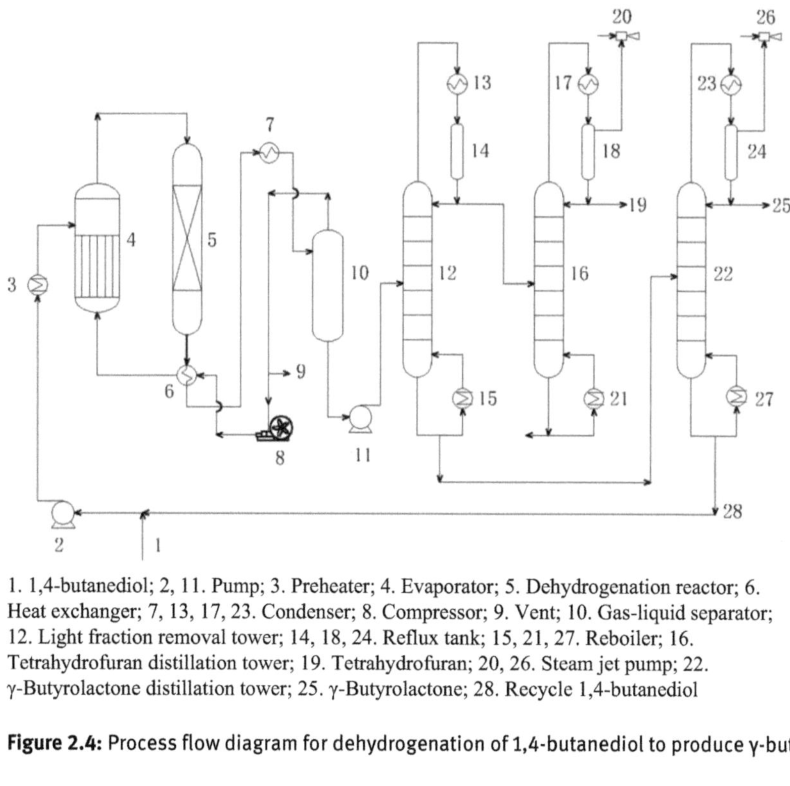

1. 1,4-butanediol; 2, 11. Pump; 3. Preheater; 4. Evaporator; 5. Dehydrogenation reactor; 6. Heat exchanger; 7, 13, 17, 23. Condenser; 8. Compressor; 9. Vent; 10. Gas-liquid separator; 12. Light fraction removal tower; 14, 18, 24. Reflux tank; 15, 21, 27. Reboiler; 16. Tetrahydrofuran distillation tower; 19. Tetrahydrofuran; 20, 26. Steam jet pump; 22. γ-Butyrolactone distillation tower; 25. γ-Butyrolactone; 28. Recycle 1,4-butanediol

Figure 2.4: Process flow diagram for dehydrogenation of 1,4-butanediol to produce γ-butyrolactone.

2.3.2.2 Hydrogenation of maleic anhydride to γ-butyrolactone

The hydrogenation of maleic anhydride can be carried out either in liquid phase or in gas phase.

In 1971, Mitsubishi Chemical Corporation in Japan built the first industrial installation for production of γ-butyrolactone on a scale of 1,000 t a^{-1} using liquid phase hydrogenation of maleic anhydride. Hokkaido Organic Chemical Factory in Japan applied a two-stage liquid phase hydrogenation process. Tetrahydrofuran Company in Japan and ICI Company in the United Kingdom adopted a one-step liquid phase hydrogenation process to build a production plant with an annual output of 1,000 tons for manufacture of BDO, γ-butyrolactone and tetrahydrofuran jointly. The ratio of these three products could be flexibly adjusted according to market demand. The process included two parts: catalytic hydrogenation reaction and crude product refining. Maleic anhydride, hydrogen and γ-butyrolactone (as a solvent) are introduced into a single-stage hydrogenation reactor containing a solid catalyst of Ni and other metals, and the reaction is carried out at a temperature of 200 °C under a pressure of 6–10 MPa.

The gas phase hydrogenation process was developed by Standard Oil Company in the United States. In China, Beijing Institute of Petroleum Science, Nankai University, and Fudan University had also carried out relatively complete research and development of this process, and successfully realized industrialization. In this process, maleic anhydride was dissolved in butanol, preheated, mixed with hot hydrogen in an evaporator and vaporized, and then the resulting gas mixture was introduced into a reactor containing a copper catalyst. The reaction was carried out at a temperature of 260–290 °C under a pressure of 0.3–0.8 MPa.

The reaction equation is

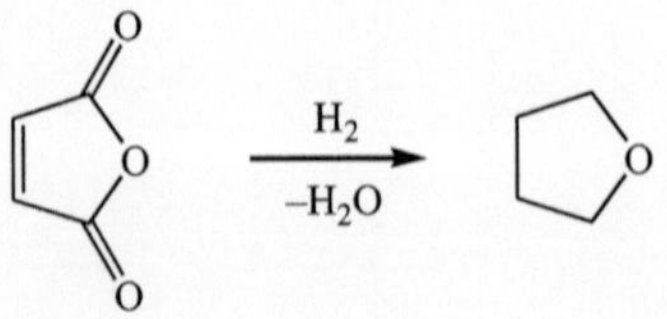

2.3.2.3 Maleic anhydride esterification

The esterification of maleic anhydride was successfully developed by Davy McKee in the United Kingdom in 1986. The process can be combined with BDO, γ-butyrolactone and tetrahydrofuran, and the ratio can be adjusted by operation conditions. The ester used in the esterification reaction may be ethyl or butyl ester. The process can be realized by adding a bypass in front of the second hydrogenation reactor of the original maleic anhydride hydrogenation process, and introducing partial product condensate of the primary hydrogenation reactor into a modified fractional distillation system for GBL. The improved rectification system coexists with the second-stage hydrogenolysis reactor, which allows the process to have higher operational flexibility and flexibility to change the proportion of the product.

The industrialized processes include furfural process, butadiene chlorination, butadiene acetoxylation, allyl alcohol process, Reppe process, maleic anhydride esterification and maleic anhydride hydrogenation. Among them, Reppe process, maleic anhydride esterification and maleic anhydride hydrogenation are more competitive. The key to the hydrogenation of maleic anhydride is the choice of catalyst. The biggest advantage is that the source of maleic anhydride is sufficient and the price is cheap. The biggest disadvantage is that the hydrogenation equipment has high requirements for explosion protection, and the hydrogen needs to be purified. Maleic anhydride esterification is a new process developed in recent years. The ratio of three products, BDO, γ-butyrolactone and tetrahydrofuran, can be adjusted by operation conditions, and the process flexibility is large. Hydrogenolysis uses non-precious metal catalysts. The yield is 90%, and there are industrial devices in Korea and Japan. Compared with the direct hydrogenation of maleic anhydride, the process contains esterification and hydrogenation, leading to complicated process flow. The Reppe process has been used in industrial production for a long time, having mature technology and large output. The copper catalyst is mainly used, and a promoter such as zinc, aluminum, chromium and manganese is added to improve the reactivity and selectivity. The catalyst is convenient to use, and has a long service life and low cost. The conversion of BDO and selectivity to γ-butyrolactone are both 98%. The economic competitiveness of the process depends primarily on the price and source of the raw material BDO. The biggest advantage is that it is not limited by the hydrogen source, and it produces hydrogen as a by-product. The production scale can be large or small, so it is suitable for a manufacturer that further produces various fine chemicals by using γ-butyrolactone as a raw material.

2.3.3 Uses of γ-butyrolactone

In addition to its use as a high-boiling solvent with excellent properties, γ-butyrolactone is an important fine chemical intermediate due to its active chemical properties with alicyclic lactone. It can react with a variety of compounds, especially ring-opening reactions, or reactions where the oxygen atom on the ring is replaced by other heteroatoms, which can be used to produce a variety of important and versatile fine chemical products.

2.3.3.1 For the production of *N*-methyl-2-pyrrolidone

γ-Butyrolactone undergoes an amination reaction with methylamine to produce *N*-methyl-2-pyrrolidone:

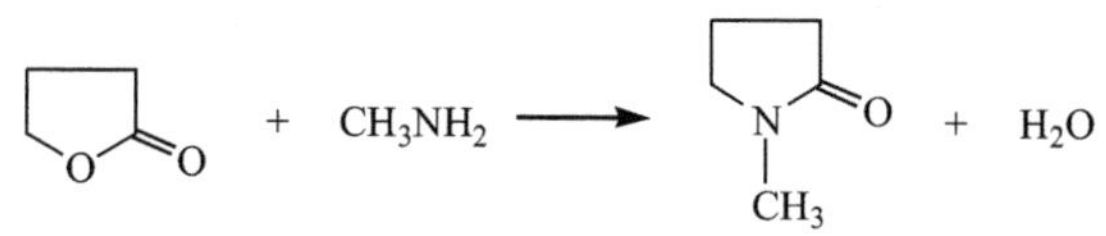

2.3.3.2 For the production of 2-pyrrolidone

Ammonolysis of γ-butyrolactone gives 2-pyrrolidone:

About 90% of 2-pyrrolidone in the world is applied for the synthesis of *N*-vinylpyrrolidone and polyvinylpyrrolidone.

2.3.3.3 For the production of cyclopropylamine

Cyclopropylamine is an important aliphatic amine containing a triatomic ring. It is mainly used as an intermediate for pharmaceuticals, pesticides and fine chemicals. In particular, the consumption of cyclopropylamine in the pharmaceutical field is very large. Using γ-butyrolactone as the starting material, the production processes include five-step process and three-step process according to the reaction steps. The following is the reaction formula of the commonly used five-step method:

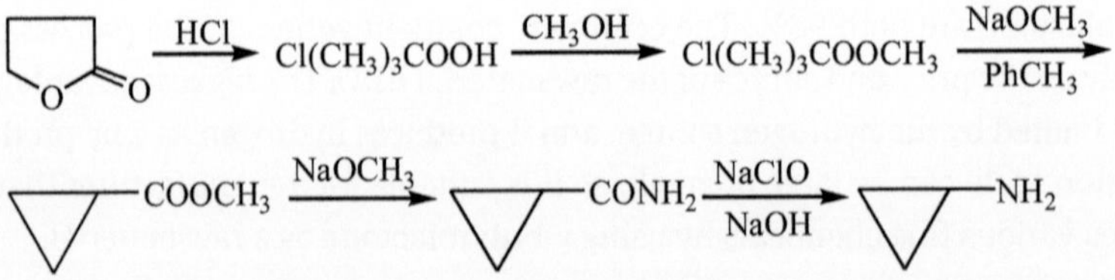

The five-step process is relatively mature and currently used by several manufacturers in China. However, the three-step process is relatively compact and advanced. Cyclopropylamine is mainly used to produce the third-generation quinolone antibacterial agent ciprofloxacin, and ciprofloxacin is one of the most active and dominant medicines in quinolone antibacterial. It has strong antibacterial activity, broad antibacterial spectrum and good absorption by the human body with little side effects. China has become one of the major producers of ciprofloxacin in the world, with an annual output of more than 1,500 tons. This alone consumes 1,000 tons of cyclopropylamine. In addition, cyclopropylamine is used in the pesticide industry to synthesize some herbicides which have broad application prospects.

2.3.3.4 For the production of γ-hydroxybutyric acid

γ-Butyrolactone is stable under neutral conditions, but under acidic conditions and high temperatures, it is hydrolyzed to produce γ-hydroxybutyric acid:

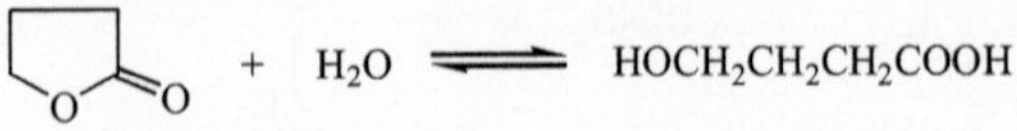

Under acidic conditions, the reaction is reversible. If it is in an alkaline environment, the rate of hydrolysis will increase and the reaction becomes irreversible.

γ-Hydroxybutyric acid has been used as a common sedative to treat insomnia, depression, narcolepsy and alcoholism. The substance is also used as an anesthetic and its use is restricted in many countries. In practice, the potassium or sodium salt of γ-hydroxybutyric acid is usually used.

2.3.3.5 For the production of succinic acid

Succinic acid can be obtained by oxidation of γ-butyrolactone:

Succinic acid is an important intermediate used in the pharmaceutical industry for the production of sedatives, diuretics, antipyretic and analgesics, contraceptives and cancer treatments. Succinic acid esters and salts can be used in synthetic dyes, alkyd resins, vinyl resins, adhesives and rubber plasticizers, ion exchange resins, various pesticides and fine chemical intermediate *N,N*-diethyl succinamide propyl ester and the like.

2.3.3.6 For the production of α-acetyl-γ-butyrolactone

γ-Butyrolactone reacted with ethyl acetate to produce α-acetyl-γ-butyrolactone:

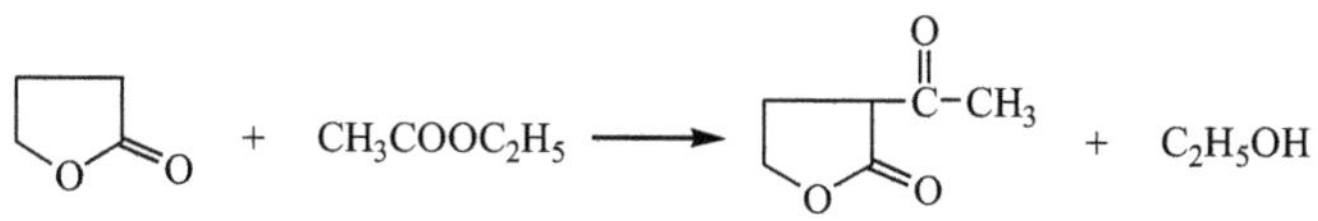

The yield of this process is slightly higher than the condensation process using ethyl acetoacetate and ethylene oxide. α-Acetyl-γ-butyrolactone is an important pharmaceutical and pesticide intermediate, mainly used in the synthesis of vitamin B_1. It is also used in the synthesis of 3,4-disubstituted pyridine and 5-(β-hydroxyethyl)-4-methylthiazole.

2.3.4 Production status of γ-butyrolactone

In 2010, the global production capacity of γ-butyrolactone was about 350,000 tons, and the output was about 300,000 tons. Production of γ-butyrolactone is mainly concentrated in the United States, European Union, Japan and China. Among them, the major US manufacturers include Lyondell and ISP. In Western Europe, the manufacturers include BASF (Germany), ICI (United Kingdom) and UCB (Belgium). In Japan, the manufacturers of γ-butyrolactone are Mitsubishi Chemical, Idemitsu Petrochemical, Hokkaido Organic Chemical and Tetrahydrofuran Company. In 2006, BASF expanded its production capacity of γ-butyrolactone in Germany up to a capacity of 85,000 t a^{-1}. It is the largest producer of γ-butyrolactone in the world.

Each country has its own focus on the consumption of γ-butyrolactone. The consumption structure of γ-butyrolactone in the United States is roughly: 56% for *N*-methylpyrrolidone (NMP) production, and 25–30% for 2-pyrrolidone, *N*-vinylpyrrolidone and polyvinylpyrrolidone production, others account for 14–19%. The consumption structure of γ-butyrolactone in Western Europe is: 45% for polyvinylpyrrolidone production, 43% for NMP production and the others account for 12%. In Japan, γ-butyrolactone of 40–45% is used as an electrolyte solution, 38% is for production of NMP and the others are used as a solvent.

In China, γ-butyrolactone production began in the 1980s. By 2003, there were more than 20 manufacturers of γ-butyrolactone with a total capacity of about 35,000 tons and an annual output of about 20,000 tons. In 2005, the total production capacity of γ-butyrolactone in China was about 57,000 tons and the annual output was about 34,500 tons. In recent years, with the rapid expansion of the downstream products such as NMP and 2-pyrrolidone, the production of γ-butyrolactone has also increased sharply.

2.4 2-Pyrrolidone

2-Pyrrolidone, also known as α-pyrrolidinone, α-pyrrolidone, γ-aminobutyric lactam, γ-butyrolactam, γ-aminobutyrolactam, butyrolactam and pyrrolidone, is a derivative of γ-butyrolactone. CAS Registry Number: 616-45-5. Molecular formula: C_4H_7NO, molecular weight: 85.1045. The chemical structure is

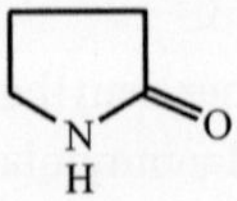

2.4.1 Properties of 2-pyrrolidone

2-Pyrrolidone is a colorless crystal at a temperature below 25 °C, yet it is a colorless to pale yellow transparent liquid at a temperature above 25 °C. Its boiling point is 245 °C under atmospheric pressure. 2-Pyrrolidone is miscible with water and most organic solvents such as alcohol, ether, chloroform, benzene, ethyl acetate and carbon disulfide but it is hardly soluble in petroleum ether. It is an excellent organic solvent with high chemical stability and high boiling point. The physical properties of 2-pyrrolidone are shown in Table 2.8.

Table 2.8: Physical properties of 2-pyrrolidone.

Melting point/°C	25	Flash point (open cup)/°C	129.4
Boiling point/°C		Heat of combustion/ kJ · mol^{-1}	2,290.6
0.133 kPa	103	Enthalpy of vaporization/ kJ · mol^{-1}	53.03
1.33 kPa	122	Enthalpy of fusion/kJ · mol^{-1}	13.92
1.60 kPa	133	Heat capacity/J · mol^{-1} · K^{-1}	
13.3 kPa	181	$C_{p,liquid}$ (300 K)	169.6
101.3 kPa	245	$C_{p,solid}$ (290 K)	135.6

Table 2.8 (continued)

Density/(g cm^{-3})		Dielectric constant (35 °C)	27.10
25 °C	1.108	Surface tension (25 °C)/N · m^{-1}	47 × 10^{-3}
50 °C	1.087	Dipole moment/×10^{-30}C · m	13.64
100 °C	1.046	Critical temperature/°C	523
Refractive index (30 °C)	1.4840	Critical pressure/MPa	6.2
Viscosity (25 °C)/mPa · s	13.3	Thermal conductivity (25 °C)/W · m^{-1} · K^{-1}	0.1861

In an aqueous solution of a strong acid or a strong base, 2-pyrrolidone is hydrolyzed to form 4-aminobutyric acid:

In the presence of a base, 2-pyrrolidone can be ring-opened and polymerized to form polypyrrolidone:

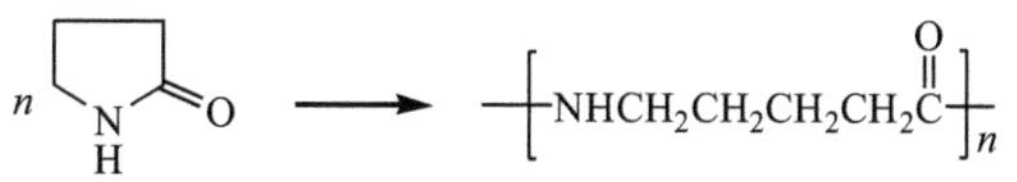

Polypyrrolidone is also known as nylon 4, which is a polymer having potential applications in textile fibers, film forming materials and molding compounds.

The hydrogen atom bonded to the nitrogen atom in the molecule of 2-pyrrolidone is relatively active and is liable to undergo a substitution reaction or a double bond addition. For example, in the presence of a base, 2-pyrrolidone can react with an alkyl halide or a dialkyl sulfate, and the hydrogen connected to the nitrogen atom is replaced by an alkyl group to produce an *N*-alkylpyrrolidone:

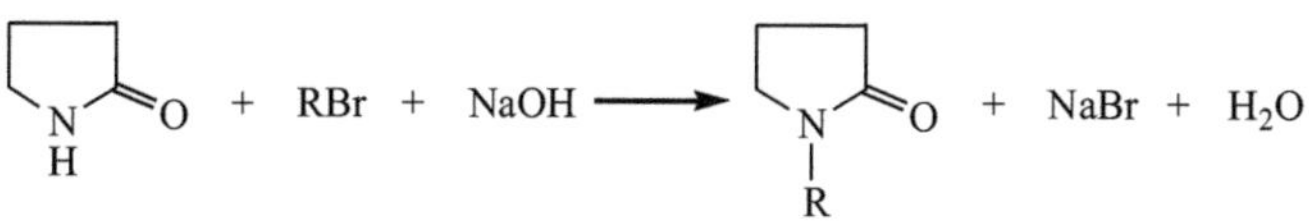

2-Pyrrolidone undergoes an addition reaction with the carbonyl group of an aldehyde to produce *N*-hydroxyalkylpyrrolidone:

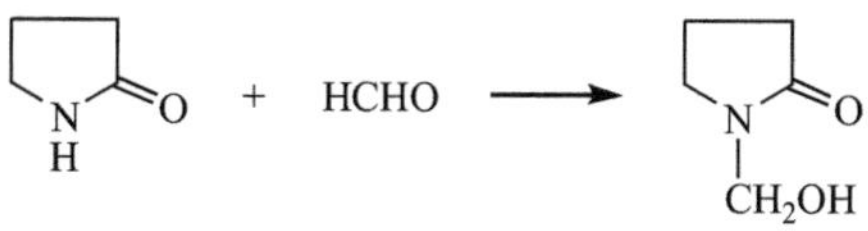

The carbonyl group of 2-pyrrolidone is hydrogenated to yield pyrrolidine in the presence of a cobalt catalyst at a high temperature:

2.4.2 Process for the production of 2-pyrrolidone

Commercial 2-pyrrolidone of industrial-grade requires a purity of more than 99.5% and a moisture content of less than 0.1%. There are several processes for production of 2-pyrrolidone in industry, including hydrogenation and amination of maleic anhydride or succinic acid, ammonolysis of γ-butyrolactone in vapor phase and ammonolysis of γ-butyrolactone in liquid phase. However, the most widely used process is the ammonolysis of γ-butyrolactone in liquid phase.

2.4.2.1 Ammonolysis of γ-butyrolactone in vapor phase

In a fixed-bed reactor packed with magnesium silicate catalyst, γ-butyrolactone and ammonia reacted in the vapor phase under the conditions including a reaction temperature of 250 °C, a pressure of 0.5 MPa, feed mass ratio of γ-butyrolactone/H_2O/NH_3 of 1:1:2 and a liquid hourly space velocity of 0.01. The conversion of γ-butyrolactone was 97% and the selectivity to 2-pyrrolidone was 96%, resulting in a yield of 93.1%.

Addition of water to the reaction system significantly improved the selectivity to 2-pyrrolidone. If the amination reaction was carried out in the absence of water, the selectivity to 2-pyrrolidone was 90% or less. However, the selectivity was enhanced to 96% by adding water of the same mass as γ-butyrolactone [10].

The magnesium silicate catalyst used in the vapor phase ammonolysis consists of 14.5 wt% MgO and 65–70 wt% SiO_2, having a surface area of 400–550 $m^2\ g^{-1}$.

Although the vapor phase ammonolysis has advantages of high selectivity and low operating pressure, it also has disadvantages of low space-time yield and a necessity of catalyst. Moreover, the feed molar ratio of NH_3/γ-butyrolactone is about 10: 1, so that a very large amount of excess and unreacted ammonia needs to be recovered and recycled.

2.4.2.2 Ammonolysis of γ-butyrolactone in liquid phase

γ-Butyrolactone and liquid ammonia (or aqueous ammonia) are reacted in the liquid phase at a temperature of 250 to 290 °C and a pressure of 8.0 to 16.0 MPa to produce 2-pyrrolidone with high conversion and selectivity:

The reaction is carried out in the absence of catalyst. In general, a tubular reactor operated in continuous mode is used in industrial production of 2-pyrrolidone, as shown in Figure 2.5, yet an autoclave can also be used for batch preparation on a small scale.

The molar ratio of γ-butyrolactone/NH_3/H_2O is 1: (2.2 to 3): (1.6 to 2.3). The residence time of the reaction mixture in the tubular reactor is usually from 20 to 120 min. The conversion of γ-butyrolactone is close to 100%, and the selectivity to 2-pyrrolidone is higher than 94%. The presence of water in the reaction system improves the selectivity, as it does in the process of vapor phase ammonolysis [11].

The excess and unreacted ammonia is distilled off in a distillation tower and sent back to an absorption tower to be absorbed to a certain concentration for recycling to the reaction system.

Due to a considerable energy loss by the jacket heating of the tubular reactor, in order to save energy, some manufacturers have modified the tubular reactor. They use a tubular reactor without the jacket and immerse the whole reactor in the constant temperature heat transfer oil tank to maintain the reaction temperature instead of using the jacket heating. The energy saving is significant after the transformation.

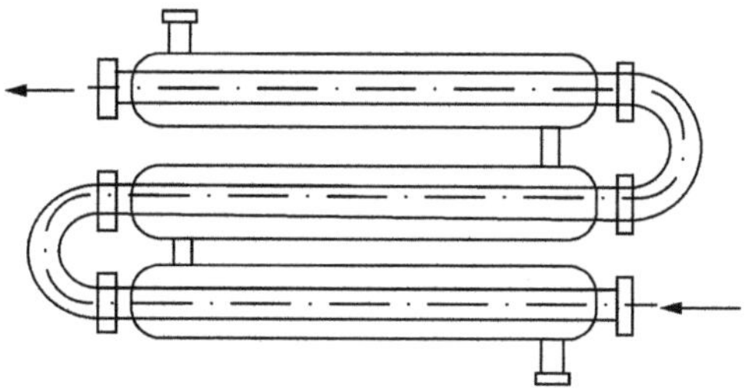

Figure 2.5: Tubular reactor.

2.4.2.3 Hydrogenation and amination of maleic anhydride or succinic acid

The ammonia aqueous solution of maleic anhydride is used as a raw material and the process is actually a hydrogenation and amination of butenedioic acid, succinic anhydride or succinic acid. Generally, activated carbon-supported noble metals such as ruthenium, rhodium and palladium are used as catalysts, and the precious metal loading is generally about 5%. A slightly less active Raney nickel catalyst can also be used. Generally, the molar ratio of ammonia to maleic anhydride is 2–5 : 1, the reaction temperature is 200–300 °C, the pressure is 5–15 MPa and the reaction time is 3–5 h. The conversion of maleic anhydride or succinic acid and the yield of 2-pyrrolidone are both 90% or more. The reaction equation is as follows:

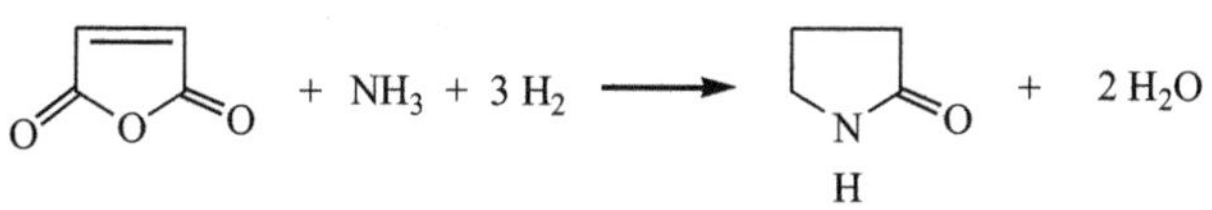

2.4.3 Uses of 2-pyrrolidone

2.4.3.1 Application for production of *N*-vinylpyrrolidone and polyvinylpyrrolidone

The most important use of 2-pyrrolidone is as a raw material for the production of *N*-vinylpyrrolidone and polyvinylpyrrolidone, which currently accounts for about 90% of the total production of 2-pyrrolidone in the world.

2.4.3.2 Use as a solvent and extractant

2-Pyrrolidone is a high-boiling polar solvent which is miscible with water, alcohol, ether, ester, ketone, benzene, carbon tetrachloride, chloroform, carbon disulfide and so on. It can also be used as a solvent in the production of synthetic resin, pesticides, insecticides, polyols and ink. Furthermore, it is also used as a solvent for synthesis of vitamin B1, myleran and the like. Moreover, 2-pyrrolidone can be used as an extractant for separation of aromatic hydrocarbons.

2.4.3.3 Use as a pharmaceutical intermediate

2-Pyrrolidone is hydrolyzed under strong acid or strong base conditions to form aminobutyric acid, which can be used as a nutrient for the brain and nervous system and as a supplement to food. 2-Pyrrolidone reacts with sodium methoxide, then the resulting sodium salt of 2-pyrrolidone undergoes a condensation reaction with ethyl chloroacetate and an amination reaction with ammonia to obtain pyrrolidone acetamide, which is also a nervous system nutrition medicine that enhances brain memory and slows memory decline in the elderly people.

2.4.4 Production status of 2-pyrrolidone

In 2014, the global production output of 2-pyrrolidone exceeded 60,000 t a^{-1}, including 25,000–30,000 t a^{-1} in the United States and 10,000 t a^{-1} in China. The major producers in the world are BASF (Germany) and Ashland (United States). The main producers in China are CNSG Anhui Hong Sifang Co., Ltd., Shandong Jiuheng Group Co., Ltd., Shandong Xintai Chemical General Plant, Shanghai Yuanji Chemical Co., Ltd., Henan Wei, Weihui Yubei Chemical Factory, Taizhou Liansheng Chemical Co., Ltd., Northeast Pharmaceutical Group Corporation, Nanjing Jinlong Chemical Factory and so on. Most of the manufacturers of polyvinylpyrrolidone have their own installations for the production of 2-pyrrolidone, which is the key intermediate for the monomer *N*-vinylpyrrolidone.

2.5 *N*-Methylpyrrolidone

NMP, also known as *N*-methyl-α-pyrrolidinone; *N*-methyl-α-pyrrolidone, *N*-methyl-γ-butyrolactam, *N*-methyl-2-pyrrolidinone, *N*-methyl-2-pyrrolidone, *n*-methylpyrrolidinone, 1-methyl-2-pyrrolidinone and 1-methyl-2-pyrrolidone, is a derivative of γ-butyrolactone. CAS registry number: 872-50-4. Molecular formula: C_5H_9NO, molecular weight: 99.1311. The chemical structure is

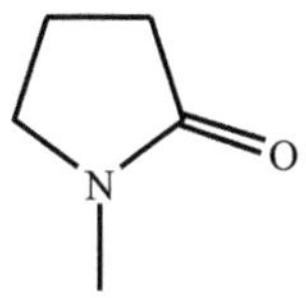

2.5.1 Properties of NMP

NMP is a colorless and transparent liquid with a slight ammonia smell. It is a high-polarity, chemically and thermally stable high-boiling solvent (boiling point: 202 °C). It is completely miscible with water, alcohol, ether, ketone, ethyl acetate, chloroform and benzene. The solubility of unsaturated hydrocarbons such as alkynes, olefins and diolefins in NMP is higher than that of saturated hydrocarbons. In addition, a variety of polymers can also be dissolved in NMP. The physical properties of NMP are shown in Table 2.9.

Table 2.9: Physical properties of NMP.

Melting point/°C	−24.4	Flash point (open cup)/°C	95
Boiling point/°C	202	Heat of combustion /$kJ \cdot mol^{-1}$	3,010
Vapor pressure/kPa		Enthalpy of vaporization/ $kJ \cdot mol^{-1}$	
41 °C	0.133	77 °C	53.4
60 °C	0.459	122 °C	49.5
79 °C	1.33	152 °C	47.7
100 °C	3.20	Enthalpy of fusion/$kJ \cdot mol^{-1}$	18.1
136 °C	13.3	Heat capacity (25 °C)/$J \cdot mol^{-1} \cdot K^{-1}$	412.4
150 °C	21.6	Critical temperature/°C	448.5
Density (25 °C)/$g \cdot cm^{-3}$	1.028	Critical pressure/MPa	4.52
Refractive index (25 °C)	1.4680	Dipole moment/ $\times 10^{-30}$ $C \cdot m$	13.64
Viscosity (25 °C)/$mP \cdot s$	1.65	Dielectric constant (25 °C)	32.10
Surface tension (25 °C)/$N \cdot m^{-1}$	41×10^{-3}	pH (10%)	7.9

NMP shows weak basicity and the pH of its 10% aqueous solution is 7.9. It is relatively stable in neutral solution, even if heated to 200 °C, the hydrolysis of NMP is only about

2.5% [12]. However, it is easy to be hydrolyzed in the presence of strong acid or strong base. The product of hydrolysis is mainly 4-methylaminobutyric acid. Oligomers from 4-methylaminobutyric acid may also be formed. In an aqueous solution of concentrated hydrochloric acid, NMP is gradually hydrolyzed to form 4-methylaminobutyric acid:

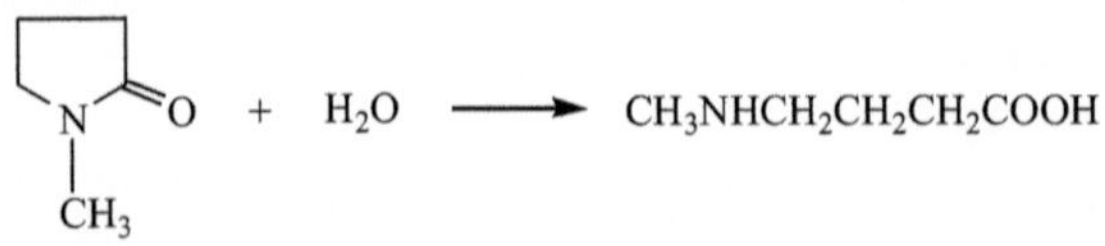

2.5.2 Process for manufacture of NMP

NMP is a low toxicity and highly efficient selective solvent which is widely used in aromatics extraction, acetylene enrichment, lubricating oil refining, extraction of butadiene and isoprene, and absorbent for natural gas or syngas purification. Based on the raw materials used, several processes such as succinonitrile route, condensation of γ-butyrolactone and methylamine, 4-oxidized methyl butyrate method and 1,4-succinic acid route for manufacture of NMP were developed. However, the only process currently employed in the industry is the condensation from γ-butyrolactone and methylamine [13]. The process technologies are similar in all NMP producers, yet the main difference is the production method for γ-butyrolactone intermediate. BASF developed a process included a dehydrogenation of BDO to γ-butyrolactone, and then a condensation with methylamine to produce NMP. Mitsubishi Chemical Corporation of Japan developed a process included a hydrogenation of maleic anhydride to γ-butyrolactone, and then condensation with methylamine to yield NMP.

Synthesis of NMP by the condensation of γ-butyrolactone and methylamine in the absence of a catalyst was developed by BASF in 1936. γ-Butyrolactone and methylamine are reacted at a temperature of 240–285 °C and a pressure of 5.0–8.0 MPa for 2–3 h, and the resulting NMP is purified by fractional distillation to obtain a high-purity product. Its reaction equation is as follows:

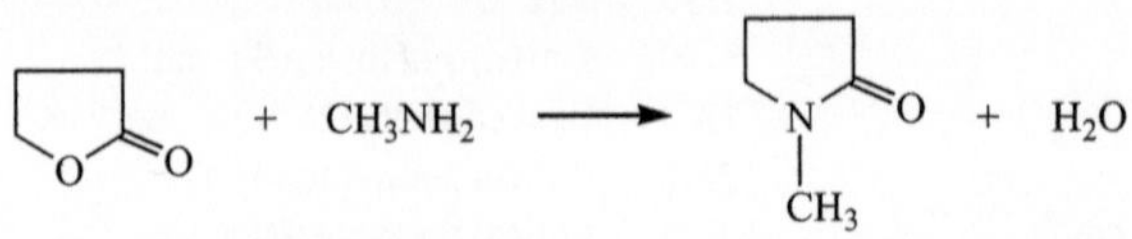

The reaction is carried out in two steps. The first step is the reaction of γ-butyrolactone with methylamine to form *N*-methyl-γ-hydroxybutanamide, which is a reversible reaction and can be carried out at lower temperature and lower pressure. The second step is the dehydration and cyclization of *N*-methyl-γ-hydroxybutanamide to NMP under higher temperature and higher pressure. The two-step reactions can be carried out batchwise in an autoclave or continuously in a tubular reactor.

Since the boiling point of NMP (204 °C) is very close to that of the raw material γ-butyrolactone (206 °C), it is quite difficult to separate them from each other by fractional distillation. Therefore, large excess of methylamine is applied in the reaction (molar ratio of methylamine to γ-butyrolactone from 1.5 to 1.8) for complete conversion of γ-butyrolactone, thereby simplifying the product separation process. Anhydrous methylamine can be used as the raw material, but addition of some water in the reaction system can increase the selectivity to NMP. The reaction solution is subjected to fractional distillation in the amine removal tower, in which the unreacted methylamine is distilled off from the top and recycled to the reactor. The bottom product of the tower is introduced into the light fraction removal tower, and all the remaining water is distilled off from the top of the tower, and the light fraction is collected from the side line on the upper section of the tower. The bottom product of the light fraction removal tower is introduced into the heavy fraction removal tower. The heavy fraction is removed from the bottom of the tower, and NMP of high purity is obtained from the top of the tower. The yield of NMP is up to 99% based on γ-butyrolactone. The process flow diagram is shown in Figure 2.6.

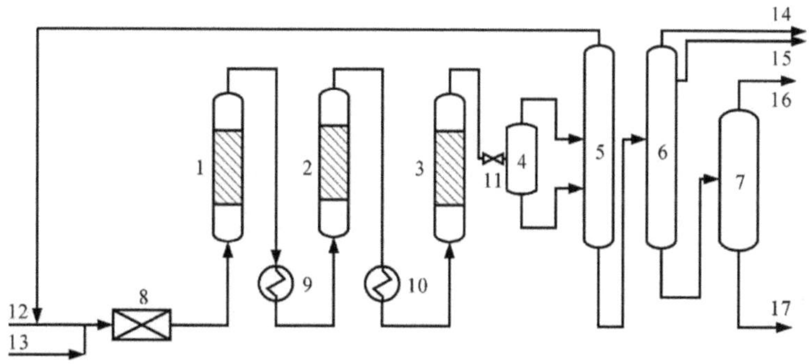

1. Amination reactor; 2. Cyclization reactor; 3. Subsequent cyclization reactor; 4. Flash tank; 5. Amine removal tower; 6. Light fraction removal tower; 7. Heavy fraction removal tower; 8. Static mixer; 9, Heater 1; 10. Heater 2; 11. Throttle valve; 12. Methylamine; 13. γ-Butyrolactone; 14. Water; 15. Light byproducts; 16. NMP; 17. Heavy byproduct.

Figure 2.6: Process flow diagram for continuous and non-catalytic synthesis of NMP from γ-butyrolactone and monomethylamine.

The process for manufacture of NMP from γ-butyrolactone and methylamine in the absence of a catalyst requires higher reaction temperature, pressure and longer residence time, which results in higher requirements on equipment selection, process operation and control, and thus increases the investment. If a catalyst is used, the reaction can be carried out under a lower pressure, and the equipment investment will be reduced [14].

The production of NMP is often accompanied by the production of its raw material γ-butyrolactone. In terms of cost, the further upstream raw materials used, the lower the costs of intermediate links, including labor, transportation and engineering costs. In addition, if BDO is applied as the starting material, the dehydrogenation intermediate γ-butyrolactone can be aminated with methylamine without purification, thus the corresponding operation steps can be reduced, and the energy consumption is reduced. Similarly, if a methylamine production unit is provided, separation of a mixture of methylamine, dimethylamine and trimethylamine can be eliminated, leading to a reduction of production cost. Process technology for production of NMP using mixture of methylamine, dimethylamine and trimethylamine as raw materials has been reported. Methylamine is prepared by catalytic amination of methanol with ammonia, and the resulting product is a mixture of methylamine, dimethylamine and trimethylamine, which are separated from each other by fractional distillation. Using the mixture in place of pure methylamine for the production of NMP, the cost of separation is saved. Due to the chemical reaction equilibrium of each component in the mixture, after the reaction of methylamine with γ-butyrolactone, the remaining dimethylamine and trimethylamine are recycled back to the methanol amination reactor, and further converted into methylamine by reacting with ammonia.

2.5.3 Uses of NMP

NMP is a polar aprotic solvent with high boiling point, strong polarity, low viscosity, good solubility, non-corrosivity, low toxicity, strong biodegradability, low volatility, and excellent chemical and thermal stability. It is mainly used in petrochemical, plastics industry, pharmaceuticals, pesticides, dye and lithium-ion battery industries. For example, NMP is widely used in extraction of aromatic hydrocarbon, purification of acetylene, olefins and diolefins, as well as in solvent for polymers and polymerization including polyamide, polyimide, polyphenylene sulfide and other engineering plastics or aramid fiber. In addition, it is also used in manufacture of insulation materials, pesticides, pigments and electronic products [15–17].

2.5.3.1 Acetylene enrichment

Natural gas and pyrolysis gas from light naphtha contain 8% to 10% (volume) of acetylene. Recovery and enrichment of the acetylene from these gases by NMP can increase the purity of acetylene to 99.7%. The acetylene recovery and enrichment by NMP have been used in more than 13 sets of cracking units in the world. In China, NMP has been applied to recover and enrich acetylene from natural gas in Sichuan Vinylon Factory. It is currently the largest Chinese user of NMP in acetylene recovery and enrichment, with an annual consumption of 200 tons. In recent years, many Chinese companies, including Shandong Shengli Oilfield, Henan Zhongyuan Oilfield, Sichuan Petroleum Bureau and Hainan Marine Natural Gas Company, are implementing projects for production of vinyl acetate from acetylene in natural gas, leading to a great increase in demand for NMP.

2.5.3.2 Recovery of butadiene and isoprene

The by-product C4 fraction from cracking of petroleum to ethylene at high temperature contains a large quantity of olefins. The recovery of butadiene from the C4 fraction is carried out by an acetonitrile method or a NMP process. Compared with the acetonitrile method, the NMP process is not only less toxic and noncorrosive, but also more efficient. Furthermore, the procedure for removal of alkynes by hydrogenation of the raffinate is eliminated, resulting in an energy saving of 50%. At present, the consumption of NMP for the extraction of butadiene in Xinjiang Dushanzi ethylene project is about 100 t a^{-1}. With the completion of several large-scale ethylene installations in China, NMP will be used more and more in the butadiene extraction section to recover butadiene.

Isoprene can be recovered to a purity of 99% from the crude C5 cracking fraction containing cyclopentadiene by using NMP. It has been reported that the recovery of isoprene by the NMP process in the world has accounted for more than 10% of the total capacity. In China, the application of NMP in this area started late, but now several companies have adopted the NMP process to recover isoprene.

2.5.3.3 Recovery vinyl chloride from exhaust gas in the production of polyvinyl chloride

The exhaust gas in the production of polyvinyl chloride contains a significant amount of vinyl chloride. If it is discharged into the atmosphere, it will not only waste resources, but also pollute the environment and endanger health. Tianjin Dagu Chemical Factory in China conducted a pilot test for the recovery of vinyl chloride from the exhaust gas in polyvinyl chloride production. Comparing various solvents, it is suggested that NMP is the best solvent for pressure swing absorption–desorption cycle. This technology can be extended and applied to many manufacturers of polyvinyl chloride.

2.5.3.4 Refinement of lubricating oil

The lubricating oil obtained from crude oil contains some substances that are harmful to the performance. These substances mainly include aromatic hydrocarbons and some compounds containing sulfur, nitrogen and oxygen. The presence of these harmful substances lowers the viscosity index of the lubricating oil, deteriorates the oxidation stability and deepens the color of the lubricating oil. Furthermore, they are oxidized to produce precipitation and acidic substances, resulting in oil path blockage and equipment corrosion. The refining of the lubricating oil is a process for removing the above-mentioned harmful substances and making the antioxidant stability, viscosity-temperature characteristics, residual carbon value and color of the lubricating oil to conform to the product specification standards. Solvent extraction is the most widely used method for the refining of lubricating oil. The traditional refining processes undergo extractions by phenol or furfural. Exxon has developed a refining process using NMP as the extractant. Compared with phenol or furfural, NMP has stronger solubility, higher selectivity, better chemical and thermal stability, higher boiling point, and lower toxicity. The NMP refining process is safer to operate, with a lower rate of extractant loss and lower energy consumption. NMP refining is replacing phenol or furfural refining processes, and now NMP has been used as the extractant in dozens of installations worldwide to refine lubricating oil.

2.5.3.5 Removal of acid gases

NMP is a solvent with weak alkaline. It is applied to remove acid gases (hydrogen sulfide, carbon dioxide, organic sulfides, etc.) from the synthesis and natural gases.

2.5.3.6 Solvent for polymer and polymerization

Resins are soluble in NMP with high solubility. NMP is used as a resin processing solvent in the manufacture of vinyl coatings, floor coatings or varnishes and composite coatings. NMP is used as a solvent for the polymerization reaction, for example, after reaction of benzenetetracarboxylic acid diester with diamine to form a polyamide compound, a cyclization reaction is carried out at a temperature below 50 °C in NMP solvent to obtain a polyimide. Polyimine or polyamine-polyimine heat-resistant resins and fibers are rapidly increasing, and such resins are also used in the production of high temperature cable coatings. This is a big potential market for NMP.

2.5.3.7 Other uses

In the production of fine chemicals such as pharmaceuticals, pesticides, pigments and cleaning agents, NMP also has a certain amount of consumption, but the proportion is very small. NMP is a very important auxiliary material for the production of lithium ion battery electrodes because of its unique non-toxicity, high boiling point, strong polarity, low viscosity, low corrosion, high solubility, low volatility and good stability, which is most commonly used in the pre-batching process of lithium ion batteries. It plays a very important role in the production process of lithium batteries.

2.5.4 Status of NMP production

In 2007, the global total capacity of NMP production reached 130,000 tons. In 2010, the global total capacity of NMP production was increased to 160,000 tons, and the actual output was between 130,000 and 150,000 tons. Ashland (United States), BASF (Germany), Mitsubishi Chemical Corporation (Japan), and Hokkaido Organic Chemicals Co., Ltd. (Japan) are major manufacturers of NMP. The production of NMP is growing at an average annual rate of about 8%.

In China, the development and production of NMP was relatively late. Until 1992, after the development of γ-butyrolactone production achieved great progress, the production of NMP was developed. In 1994, NMP production capacity was less than 2,000 t a^{-1}. At that time, it mainly relied on imports to meet the demand. In recent years, the production of NMP in China has developed rapidly, and several NMP production installations have been built and put into operation. In 2005, NMP production capacity reached 25,000 t a^{-1}, and in 2010 it reached 50,000 t a^{-1}, with an average annual growth rate of 20% in 5 years. In recent years, with its large-scale application in the lithium battery industry, the production capacity of NMP is increased sharply. At present, there are many manufacturers of NMP in China, but the production capacity of individual enterprise is relatively small. The main manufacturers are Nanjing Jinlong Chemical Factory (4,000 t a^{-1}) and Zhejiang Liansheng Chemical Industry Co., Ltd. (10,000 t a^{-1}), Maiqi Chemical Co., Ltd. (52,000 t a^{-1}), Taizhou Yanling Fine Chemical Co., Ltd. (9,000 t a^{-1}), Chongqing Zhongren Chemical Co., Ltd. (60,000 t a^{-1}). Moreover, other producers include Sichuan Tianhua Fubang Co., Ltd., Hunan Nanling Chemical Co., Ltd., Shandong Qingyun Changxin Chemical Co., Ltd., Shengli Oilfield Dongsheng Xingrun Chemical Co., Ltd., Jiaozuo Zhongwei Chemical Co., Ltd. and Nanjing Ruize Fine Chemical Co., Ltd.

2.6 2-Pyrrolidone acetamide

2-Pyrrolidone acetamide, also known as pirazetam, pirroxil, pyracetam, pyramem and 2-pyrrolidinone acetamide, is a derivative of 2-pyrrolidone. CAS registry number: 7491-74-9. Molecular formula: $C_6H_{10}N_2O_2$, molecular weight: 142.1558. The chemical structure is

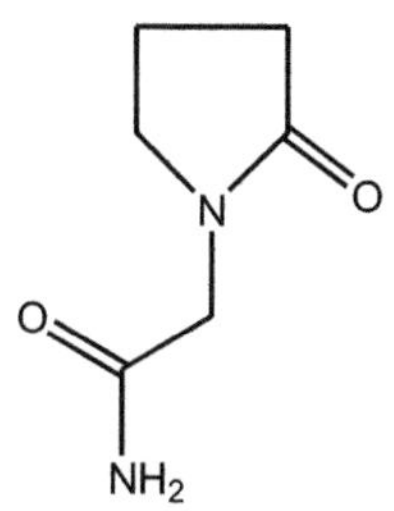

2.6.1 Properties of 2-pyrrolidone acetamide

2-Pyrrolidone acetamide is a white crystalline powder with a melting point of 151.5 to 152.5 °C. It is odorless and tastes slightly bitter. 2-Pyrrolidone acetamide is soluble in water, slightly soluble in ethanol and almost insoluble in ether.

2.6.2 Process for manufacture of 2-pyrrolidone acetamide

A process for manufacture of 2-pyrrolidone acetamide includes three steps. Firstly, 2-pyrrolidone reacts with sodium methoxide to yield sodium salt of 2-pyrrolidone:

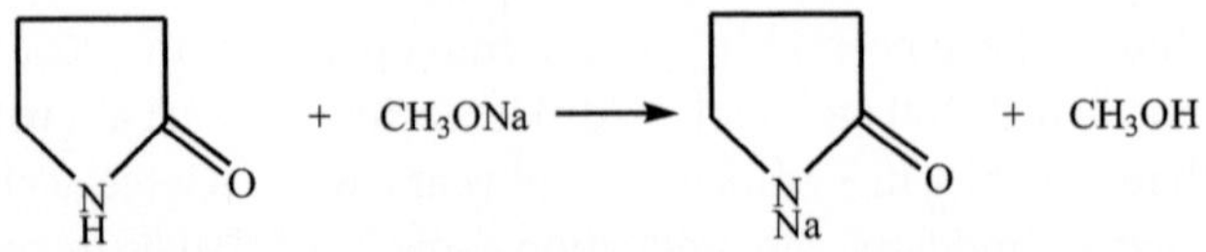

Secondly, the sodium salt of 2-pyrrolidone reacts with ethyl chloroacetate to form ethyl 2-pyrrolidoneacetate by condensation reaction:

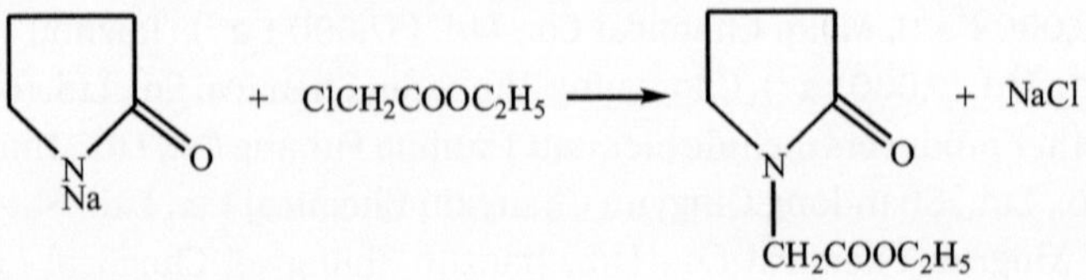

Thirdly, ethyl 2-pyrrolidoneacetate reacts with ammonia to produce 2-pyrrolidone acetamide by amination:

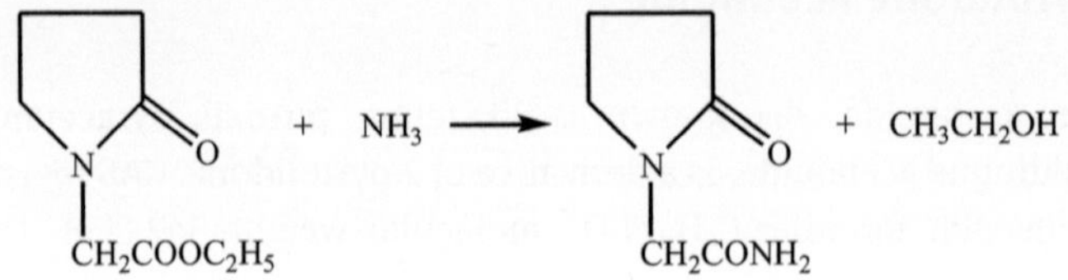

The process flow diagram is shown in Figure 2.7. About 850 kg of 2-pyrrolidone was dissolved in sodium methoxide solution to form sodium salt of 2-pyrrolidone which was soluble in water. After the crystals were dissolved in a solvent, about 128 kg of ethyl chloroacetate was added to form ethyl 2-pyrrolidoneacetate by condensation reaction. Ammonia gas was introduced, and the reaction mixture was stirred for 2 h. Then, the reaction mixture was allowed to stand, crystallized and dried to obtain 2-pyrrolidone acetamide. The yield was 76%.

2.6.3 Uses of 2-pyrrolidone acetamide

2-Pyrrolidone acetamide was used as an anti-vertigo and antiemetic drug in the early stage. It is currently used as a nervous system nutrition medicine to promote and enhance the memory function of the brain. It is effective for senile memory loss and effective for the intelligence improvement of some low-skill children. It also has a

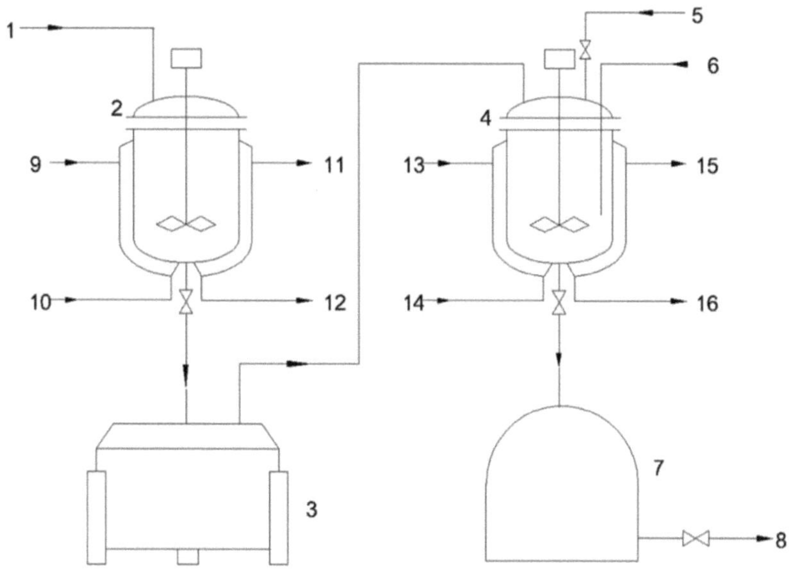

1. 2-Pyrrolidone and sodium methoxide; 2. Salt-forming reactor; 3. Centrifuge; 4. Amination reactor; 5. Ethyl chloroacetate; 6. Ammonia; 7. Product tank; 8. 2-pyrrolidone acetamide; 9, 13. Steam; 10, 14. Cooling water inlet; 11, 15. Cooling water outlet; 12, 16. Condensate drain

Figure 2.7: Process flow diagram for production of 2-pyrrolidone acetamide.

certain effect on memory and thinking disorders caused by Alzheimer's disease and brain trauma.

2.7 *N*-Vinylpyrrolidone

N-Vinylpyrrolidone, abbreviated as NVP, also known as *N*-vinyl-2-pyrrolidone, *N*-vinyl-2-pyrrolidinone, *N*-vinylpyrrolidinone, vinylbutyrolactam, vinylpyrrolidinone, vinylpyrrolidone, 1-vinyl-2-pyrrolidinone, 1-vinyl-2-pyrrolidone, 1-vinylpyrrolidinone, vinyl-2-pyrrolidone, 1-vinylpyrrolidone and 1-ethenyl-2-pyrrolidinone, is a derivative of 2-pyrrolidone and acetylene. CAS Registry Number: 88-12-0. Molecular formula: C_6H_9NO, molecular weight: 111.1418. The chemical structure is

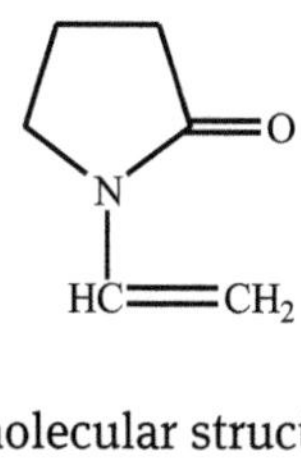

It is not difficult to find from the molecular structure of NVP that the NVP molecular structure contains an active unsaturated vinyl group (CH_2=CH–), a hydrophilic group

(N atom and carbonyl group), and a hydrophobic group (hydrocarbyl group and the cyclic hydrocarbon group), thus it has many excellent properties.

2.7.1 Properties of NVP

2.7.1.1 Physical properties of NVP

NVP is a colorless transparent liquid at room temperature. It is completely miscible with water, alcohol, ether, ester, chlorinated hydrocarbon and aromatic hydrocarbon, yet it is only partially soluble in aliphatic hydrocarbons. The physical properties of NVP are summarized in Table 2.10.

Table 2.10: Physical properties of NVP.

Melting point/°C	13.5	Density (20 °C)/g · cm^{-3}	1.043
Boiling point/°C		Refractive index (20 °C)	1.5120
13.3 Pa	58	Viscosity (20 °C)/mP · s	2.4
133 Pa	65	Flash point (open cup)/°C	95
1.20 kPa	88	Ignition point/°C	240
1.30 kPa	90–92	Heat capacity/kJ · kg^{-1} · K^{-1}	
1.867 kPa	95	0 °C	1.70
13.33 kPa	148	50 °C	2.46
Enthalpy of fusion/kJ · mol^{-1}	15.28		

Figure 2.8 shows the solid–liquid equilibria phase diagram of NVP and 2-pyrrolidone binary system. It can be concluded that the phase diagram corresponds to simple eutectic type. The corresponding eutectic mole composition is 0.5427, and the eutectic temperature is 263.75 K. The solid–liquid equilibria data play a crucial role in the purification of NVP by melt crystallization [18].

2.7.1.2 The chemical properties of NVP

NVP has some special chemical properties because of the existence of vinyl group on its nitrogen atom. The most important of them are easy polymerization and hydrolysis.

2.7.1.2.1 Hydrolysis

Under acidic conditions, NVP is easily hydrolyzed to 2-pyrrolidone and acetaldehyde:

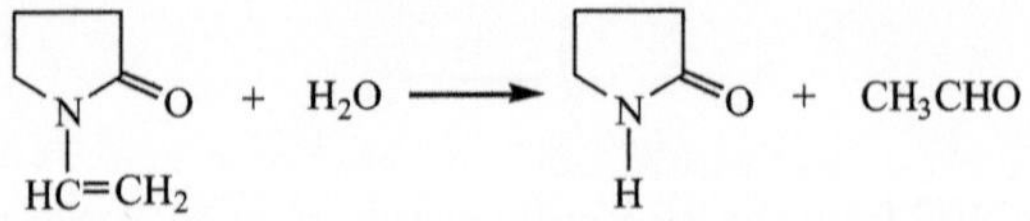

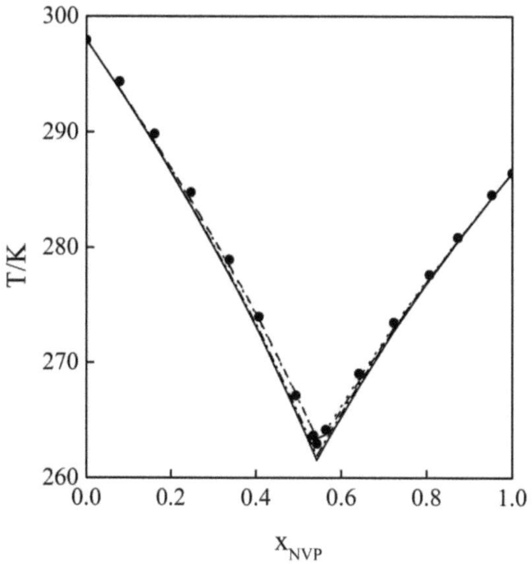

Figure 2.8: Solid–liquid equilibria phase diagram for binary system of NVP + 2-pyrrolidone: points, experimental data; solid line, the ideal solubility model; dash line, the Wilson model; dot line, the NRTL model; dash dot line, the UNIFAC model.

In order to prevent NVP products from hydrolysis during production and storage, it is necessary to remove water from NVP. In general, the water content in NVP should be less than 0.1 wt%. During storage and transportation, a 0.1% base such as sodium hydroxide, ammonia or a low molecular weight amine can be added to keep the product weakly alkaline for inhibiting hydrolysis or polymerization.

2.7.1.2.2 Polymerization

The NVP monomer readily undergoes polymerization under the action of heat or an initiator. Even in the absence of an initiator, different degrees of polymerization occur when a NVP sample is left for a long time, which has a negative impact on the quality and reactivity of the monomer. Therefore, a polymerization inhibitor is generally added to the commercially available NVP, and the inhibitor must be removed before the monomer is applied to polymerization reaction for production of polyvinylpyrrolidone and other copolymers. The polymerization inhibitor is usually removed from NVP by vacuum distillation or adsorption to obtain pure NVP.

2.7.2 Process for manufacture of NVP

The synthesis routes of NVP mainly include vinylation of 2-pyrrolidone with acetylene, amination of γ-butyrolactone with ethanolamine, and vinyl exchange of 2-pyrrolidone with vinyl acetate or vinyl ether. Among them, the production cost is the lowest for the production of NVP by vinylation of 2-pyrrolidone with acetylene, yet an acetylene generation and purification installation is required in the process. The production of NVP by amination of γ-butyrolactone with ethanolamine had been developed and adopted by some manufacturers in China during the years from 1995 to 2007. However, the cost of production is very high, even exceeding the market price of NVP with the reduction in market price of the product. Therefore, the process had been completely eliminated before 2007. As for the processes by vinyl exchange of 2-pyrrolidone with vinyl acetate or vinyl ether, they have not been applied in industry because of high costs and difficult problems of product separation from a great variety of by-products.

At present, all NVP manufacturers in the world produce NVP by vinylation of 2-pyrrolidone with acetylene.

2.7.2.1 Production of NVP by vinylation of 2-pyrrolidone with acetylene

This process is the earliest and most mature method for production of NVP, in which acetylene and 2-pyrrolidone are used as raw materials and KOH as a catalyst. The reaction mechanism is as follows:

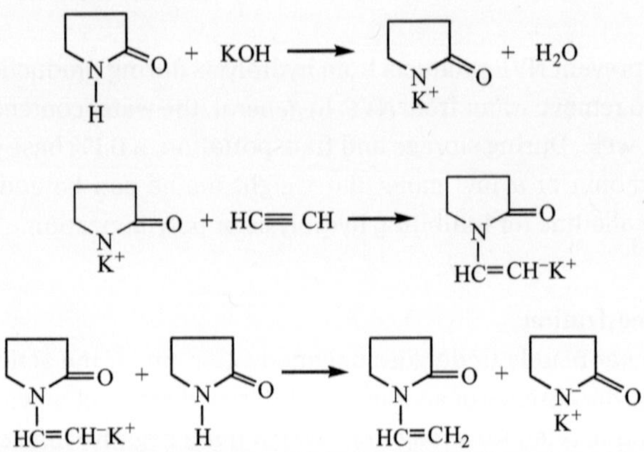

It can be seen from the reaction mechanism that the reaction is divided into two stages. The first stage is the reaction of 2-pyrrolidone with KOH to form the potassium salt of 2-pyrrolidone, which is the actually catalyst for the subsequent two-step reactions. The amount of KOH to be charged is generally from 1% to 3% by weight, and water is generated by the preparation of the potassium salt. Since the presence of water in the system causes the 2-pyrrolidone ring broken to form a by-product potassium 4-aminobutyrate, resulting in deactivation of the catalyst, it is necessary to remove the water from the system. Usually, water is removed from the solution by vacuum distillation and nitrogen bubbling. It is required to make the water content less than 0.2%. The dehydration is generally carried out at a vacuum of 730–740 mmHg and at a temperature from 80 to 130 °C.

In the second stage, acetylene is introduced and the vinylation reaction is carried out at a pressure between 0.9 and 1.5 MPa and a temperature from 150 to 180 °C. In order to ensure the safety of acetylene under pressure, it is necessary to dilute acetylene with nitrogen; that is, compressors, reactors and other equipment and pipelines are always running a mixture of acetylene and nitrogen instead of pure acetylene.

The efficiency of the reaction can be improved by replacing KOH with an alkali metal alkoxide such as potassium isobutoxide, but the cost of using potassium isobutoxide is high. It is also possible to use a crown ether or a low molecular weight polyether diol as a cocatalyst while using KOH as the catalyst. Since the crown ether has four oxygen atoms in the molecular structure ring, and the oxygen atom is more electronegative than the nitrogen atom; therefore, the oxygen atoms attract K^+ and promote K^+ to break away from the pyrrolidone ring, so that the nitrogen atom on the ring is exposed, which reduces the steric hindrance, resulting in accelerating the addition reaction with acetylene and increasing the conversion of 2-pyrrolidone. However, the use of these cocatalysts is costly and increases the difficulty of subsequent product separation.

The process for production of NVP by vinylation of 2-pyrrolidone with acetylene is shown in Figure 2.9. A solution of 2-pyrrolidone containing a homogeneous catalyst of potassium pyrrolidone is introduced into the bottom of reactor (10) via metering pump (7). A heat transfer oil for preheating the reactor is introduced into the jacket. Acetylene and nitrogen are mixed in the buffer tank (3), then the gas mixture is compressed by the acetylene compressor (4), dried by the adsorption drying columns (5) and subsequently charged to the reactor (10) from the bottom. In the reactor, 2-pyrrolidone reacts with acetylene to form NVP. The reaction mixture is discharged from the top of the reactor, cooled by the cooler (11) and introduced into the gas–liquid separator (12). After separation, the gas phase is depressurized by the pressure reducing valve (13) and then mixed with fresh acetylene feed and recycled back to the buffer tank. The liquid phase from the bottom of the gas–liquid separator is a crude

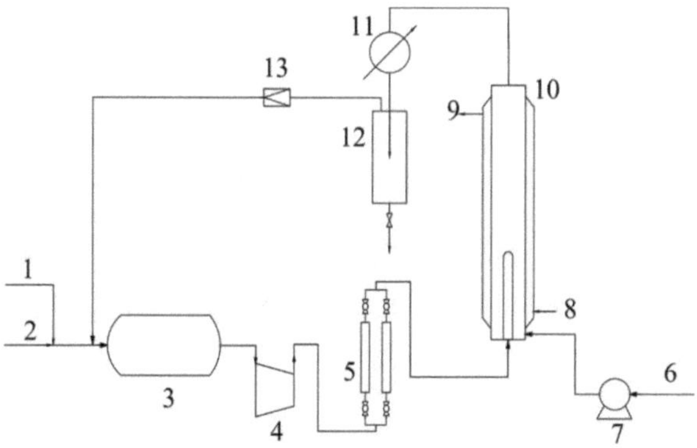

1. Nitrogen, 2. Acetylene, 3. Buffer Tank, 4. Acetylene compressor, 5. Adsorption drying column, 6. 2-Pyrrolidone, 7. Metering Pump, 8. Heat transfer oil inlet, 9.Heat transfer oil outlet, 10. Reactor, 11. Cooler,12. Gas liquid separator, 13. Pressure relief valve.

Figure 2.9: Process flow diagram of NVP production.

product containing NVP. The residence time of the reaction is from 3 to 4 h. The single pass conversion of 2-pyrrolidone is controlled between 50% and 60%, and the selectivity to NVP is above 95% [19].

The NVP crude product contains the target product NVP, unreacted 2-pyrrolidone, potassium salt 2-pyrrolidone, and small amounts of by-products. As a monomer for production of polyvinylpyrrolidone, NVP must be of high purity to be applied. The unreacted 2-pyrrolidone must also be recovered and recycled. NVP is a heat-sensitive substance. If the usual fractional distillation tower is applied for separation and purification of NVP, the material is heated all the time in the tower during the operation, which will cause a large amount of NVP to undergo heat-sensitive reactions and generate unwanted impurities, resulting in a decrease of NVP yield. Moreover, the generation of impurities also has a negative impact on the purity of the final NVP product after purification. In order to solve the problem, Li invented a method for separation and purification of NVP by using a high-vacuum batch fractional distillation with a falling film evaporator as the reboiler. Compared with the conventional tube-type reboiler, the falling film evaporator has no static pressure effect, no temperature difference loss and much smaller heat transfer temperature difference, which can make the evaporation to be carried out at a higher vacuum and a lower temperature. This technology is very suitable for the separation of heat-sensitive substances. The material containing NVP to be separated is in a cold state during the most time of an operating cycle, and the evaporation time through the falling film evaporator is very short; that is, the material is heated for a short time, thereby the heat-sensitive reactions of NVP and the formation of impurities are effectively suppressed, leading to improvements of NVP recovery and purity. The NVP rectification process is shown in Figure 2.10. The liquid mixture containing NVP to be separated in the storage tank (11) is sent to the heat exchanger (2) via the magnetic pump (1), then preheated and subsequently sent to the falling film evaporator (3). In the evaporator, the liquid mixture is heated and partially evaporated as it flows down the inner wall of the tube in a film-like manner. The organic vapor from the evaporator is introduced into the fractional distillation tower (6). The distillate from the top of the tower is condensed into a liquid by the condenser (7) and sent to the reflux tank (8). A part of the liquid is returned to the tower as a reflux liquid, and the other part is taken out and collected in the receiving tank (12 or 13). The residual liquid from the falling film evaporator and the downstream liquid from the bottom of the fractional distillation tower are cooled by the heat exchanger and returned to the storage tank. The storage tank is provided with a cooling water jacket so that the material in it is always kept close to room temperature. The reaction mixture (containing NVP, 2-pyrrolidone, catalyst and by-products) from the vinylation reactor is separated into three parts: NVP of high purity, 2-pyrrolidone containing a trace amount of NVP, and residue (catalyst and byproducts). NVP of high purity is directly used for the polymerization to produce polyvinylpyrrolidone, while 2-pyrrolidone containing a small amount of NVP is recycled for further reaction with acetylene to produce NVP [20].

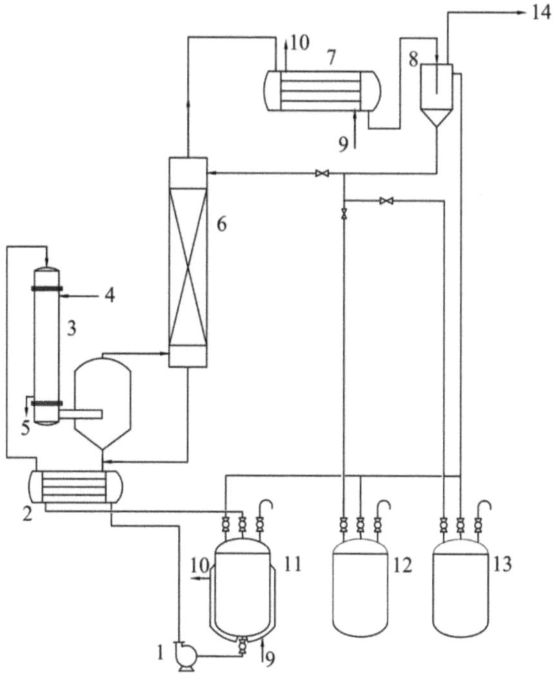

1.Magnetic pump, 2. Heat exchanger,3. Falling film evaporator,
4. Low pressure steam, 5. Steam condensate, 6. Rectification column,
7. Condenser, 8. Return tank, 9. Cooling water inlet, 10. Cooling
water outlet, 11.Tank, 12-13.Receiving tank.

Figure 2.10: Process flow diagram of NVP fractional distillation.

Similarly, a continuous high-vacuum fractional distillation system with falling film evaporators as the reboilers is shown in Figure 2.11. Compared with batch vacuum fractional distillation, continuous vacuum fractional distillation can significantly reduce steam consumption and increase the product output.

The purity of the NVP monomer obtained from fractional distillation is usually between 99.0% and 99.8%. The main impurities include 2-pyrrolidone and amine derivatives. For the polymers of NVP used in food and pharmaceutical fields, a purity of the NVP monomer higher than 99.9% is required. However, it is often very difficult to meet the requirement by fractional distillation. Moreover, the energy consumption of fractional distillation increases rapidly with the increase of product purity. To solve this problem, a melt crystallization process was applied to purify NVP from industrial grade (99.5%) to pharmaceutical grade (99.9%).

Su and Li investigated the effects of crystal seed addition amount, temperature of adding crystal seed, cooling rate, final crystallization temperature, crystallization time, heating rate, final sweating temperature and sweating time on the purity and yield of the product were investigated. The appropriate operating conditions were obtained as follows: crystal seed addition amount is of 0.1% based on the weight

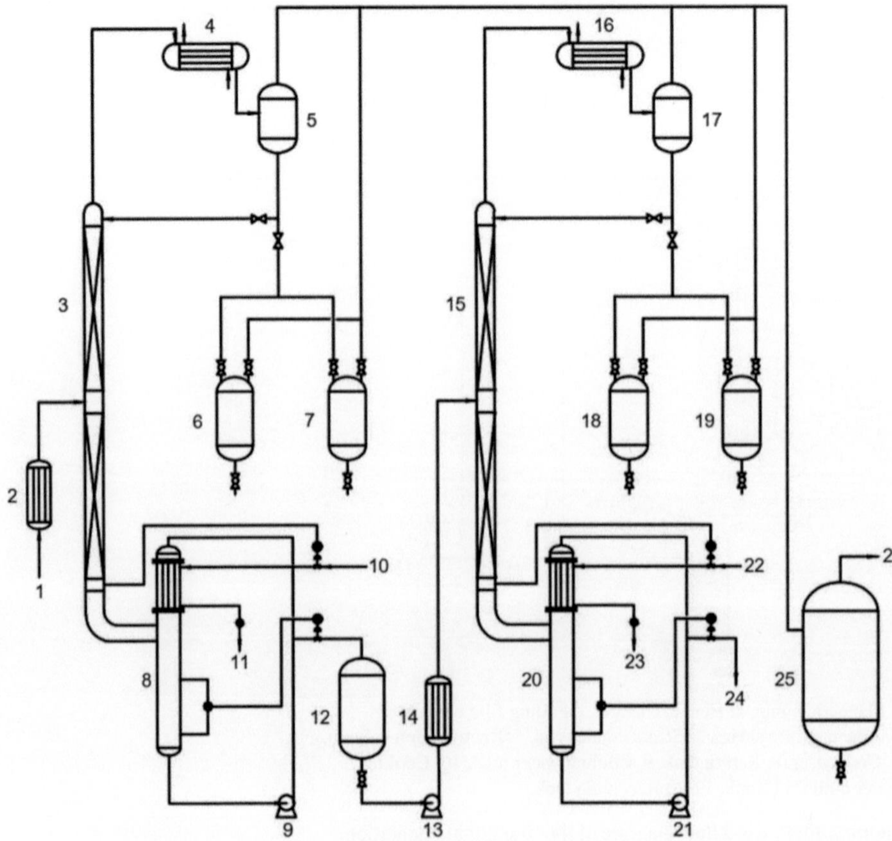

1.Feed;2, 14. Feed preheater; 3. Light fraction removal tower; 4, 16. Condenser; 5, 17. Return tank; 6,7. Receiving tanks for light fraction; 8,20. Falling filmreboiler; 9, 13, 21. Pump; 10, 22. Steam; 11,23. Steam condensate;12. Intermediate tank;15.Product distillation tower; 18, 19. Receiving tanks for heavy fraction; 24. Heavy fraction; 25. Vacuum tank; 26. Connecting to vacuum unit.

Figure 2.11: Process flow diagram for continuous fractional distillation of NVP using falling film reboiler.

of raw material, temperature of adding crystal seed of 11 °C, cooling rate of 6 °C h^{-1}, final crystallization temperature of 6 °C, crystallization time of 20 min, heating rate of 6–8 °C h^{-1}, final sweating temperature of 12 °C and sweating time of 30 min. *N*-Vinylpyrrolidone of industrial grade (99.5%) can be purified through single-stage crystallization to a purity more than 99.95% with a yield more than 74.5% [21].

Compared to other separation methods, this method has obvious advantages including only one crystallization stage, simple operation, low energy consumption, short time-consuming, good reproducibility and no requirement for any solvents.

2.7.2.2 Production of NVP by amination of γ-butyrolactone with ethanolamine

γ-Butyrolactone reacts with ethanolamine to form *N*-hydroxyethylpyrrolidone (NHP):

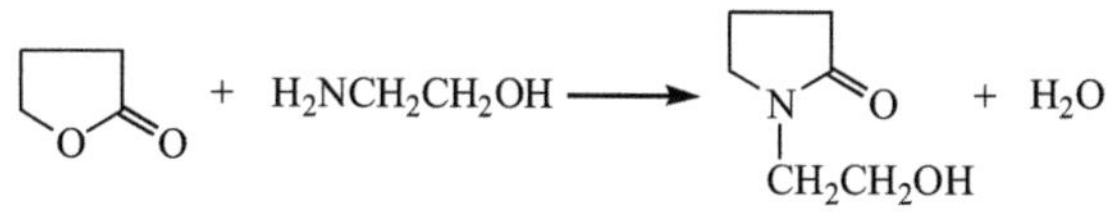

The reaction needs to be carried out at a higher temperature or in the presence of a catalyst. Certain cation exchange molecular sieves have desirable catalytic activity and selectivity for the reaction. The reaction requires an excess of ethanolamine, and the yield of NHP can reach over 90%.

NHP undergoes a dehydration reaction to form NVP. The dehydration processes include direct and indirect dehydrations.

The direct dehydration of NHP is carried out at a temperature between 300 and 340 °C in the presence of a catalyst:

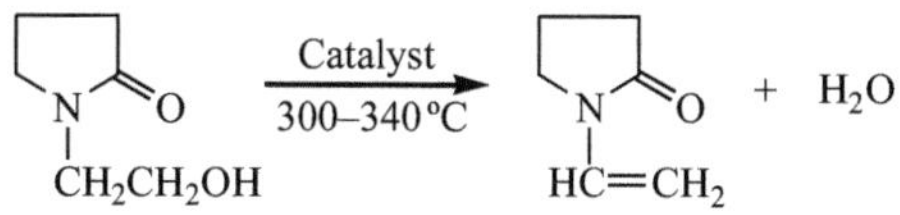

The direct dehydration of NHP is difficult, and the key is to prepare a catalyst with high activity and high selectivity. The commonly used catalysts are oxides of aluminum, zirconium, tin and silicon. The composite catalyst of zirconium oxide and tin oxide has both acid center and basic center, which has strong catalytic activity and high selectivity to NVP. At the reaction temperature of from 300 to 340 °C, the single-pass conversion of NHP is above 70%, and the selectivity to NVP is over 90%.

In the indirect dehydration process, NHP is converted into *N*-chloroethyl pyrrolidone by the reaction of NHP with halogenating agent, and then NVP is obtained by the removal of hydrogen chloride:

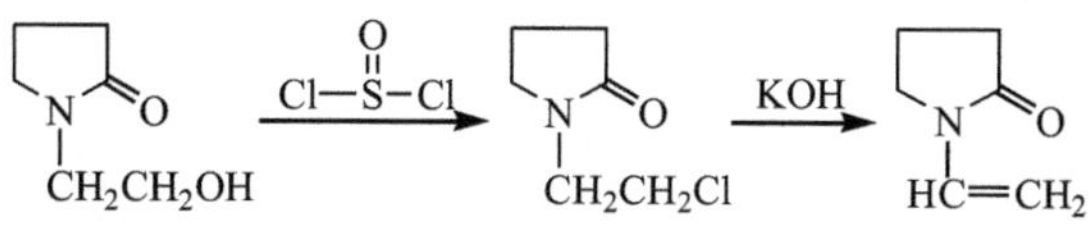

Commonly used halogenating agents are hydrogen chloride, thionyl chloride and the like. The halogenation with thionyl chloride can be carried out at room temperature in the absence of a catalyst [22].

The dehydrochlorination of *N*-chloroethyl pyrrolidone is carried out in the presence of a catalyst such as γ-Al_2O_3, KOH, $NaNH_2$ and $NaOCH_3$. The yield of NVP is about 50%.

2.7.3 Uses of NVP

N-Vinylpyrrolidone is mainly used in the synthesis of polyvinylpyrrolidone and its copolymers with other monomers. In addition, NVP is used as a low-viscosity reactive diluent monomer in the UV radiation curing process, which can accelerate the UV curing rate. It is usually compatible with acrylates and has a unique viscosity reducing effect and an enhancement of adhesion to nonpolar materials. UV-curable coatings containing NVP can be used to impart high abrasion resistance to resin lenses, glass protective films and credit cards.

2.8 Polyvinylpyrrolidone

Polyvinylpyrrolidone, abbreviated as PVP, also known as povidone, 1-ethenyl-2-pyrrolidinone homopolymer, poly[1-(2-oxopyrrolidin-1-yl)-ethylene], poly [(2-oxo-1-pyrrolidinyl)ethylene], 1-vinyl-2-pyrrolidinone polymer, poly(vinylpyrrolidinone), poly (vinylpyrrolidone), poly(1-vinyl-2-pyrrolidone), poly(*N*-vinylpyrrolidone), poly(*N*-vinyl-2-pyrrolidone), is one of the most important fine chemicals derived from acetylene. CAS Registry Number: 9003-39-8. Molecular formula: $(C_6H_9NO)_n$. The chemical structure is

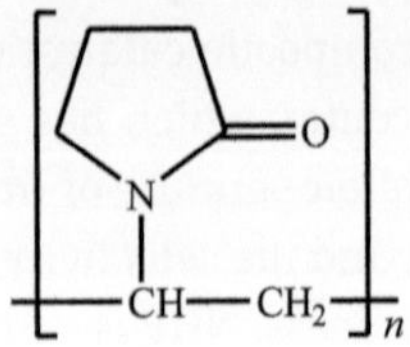

2.8.1 The properties of PVP

Commercial PVP is a white or milky white powder solid. Due to the difference in *n* number (the degree of polymerization), PVP has different molecular weights and slightly different properties. In general, PVP with different molecular weight ranges are expressed by *K*-values (e.g., *K*-12, *K*-15, *K*-17, *K*-30, *K*-60 and *K*-90). PVP is tasteless, odorless and low toxicity. It has excellent physiological inertia and biocompatibility, and has no irritation to the skin and eyes. It can be biodegraded. According to different process conditions, three grades of products including pharmaceutical, food and industrial grade are available.

In the structure of PVP, the methylene groups that form the chain and the pyrrolidone ring are nonpolar groups and have lipophilicity, and the lactam in the molecule is a strong polar group and has a hydrophilic effect. This structural feature makes PVP soluble in water and many organic solvents, such as alcohols, carboxylic acids, amines and chlorinated hydrocarbons. PVP can also be miscible with many water-soluble and water-insoluble polymers, such as dextrin, carboxymethyl cellulose, gum arabic, polyvinyl alcohol and polyethylene glycol. In addition, PVP can also be miscible with plasticizers such as phthalates, sorbitol and glycerin, so these substances can be used as plasticizers for PVP.

2.8.1.1 *K*-value and molecular weight of PVP

The average molecular weight of a polymer can be characterized in three different ways including weight average, number average and viscosity average. The weight average molecular weight, $\bar{M}_w$, is determined by light scattering, ultracentrifuge or gel permeation chromatography. The number average molecular weight, $\bar{M}_n$, is measured by osmometry or membrane filtration. The viscosity average molecular weight, $\bar{M}_v$, is determined by viscosity.

In practice, the molecular weight of commercial PVP is generally expressed by the *K*-value of the Fikentscher's method. The aqueous solution containing predetermined amount of PVP in 100 mL is mixed and kept uniform for 1 h. The relative viscosity of the aqueous solution relative to water is measured using a capillary-tube viscometer at a constant temperature water bath temperature of 25 °C, and then the *K*-value is calculated according to the Fikentscher's formula:

$$K = \frac{\left[\sqrt{300C\log z + \left(C + 1.5C\log z\right)^2} + 1.5C\log z - C\right]}{\left(0.15C + 0.003C^2\right)}$$

where C is the weight, on the anhydrous basis, of the specimen tested in each 100.0 mL of solution (g). z is the relative viscosity (viscosity of the sample solution relative to that of water) and calculated as follows:

$$z = \frac{\text{Flow time of the sample solution}}{\text{Flow time of water}}$$

The predetermined amount of PVP is on the anhydrous basis and specified in Table 2.11.

Table 2.11: The amount of PVP for *K*-value measurements.

Nominal *K*-value	Quantity (g)
≤18	5.00
>18 to ≤95	1.00
>95	0.10

For a commercial PVP product, the actual *K*-value is allowed within a certain range above and below its nominal *K*-value. According to the pharmacopoeias valid in the United States and European Union, *K*-value ranges specified for PVP are as follows: if the *K*-value of PVP having a stated (nominal) *K*-value of not more than 15, the *K*-value is allowed to be 85.0–115.0% of the stated value, while for nominal *K*-value above 15, the K-value is between 90.0% and 108.0% of the stated value. Table 2.12 gives the relationship between the *K*-value ranges and the relative viscosity values of PVP in water for PVP *K*-12, PVP *K*-15, PVP *K*-17, PVP *K*-25, PVP *K*-30, PVP *K*-60 and PVP *K*-90.

Table 2.12: Relationship between the *K*-value ranges and the relative viscosity values of PVP in water (capillary viscometer, 25 °C).

PVP grade	*K*-value	Concentration (g 100 mL^{-1})	Relative viscosity
PVP *K*-12	10.2–13.8	5	1.222–1.361
PVP *K*-15	12.75–17.25	5	1.317–1.531
PVP *K*-17	15.3–18.36	5	1.430–1.596
PVP *K*-25	22.5–27.0	1	1.146–1.201
PVP *K*-30	27.0–32.4	1	1.201–1.281
PVP *K*-60	54.0–64.8	1	1.810–2.248
PVP *K*-90	81.0–97.2	1	3.310–5.195

If the *K*-value of PVP has been determined, the viscosity average molecular weight $\bar{M}_v$ can be calculated as follows:

First, calculating the intrinsic viscosity $[\eta]$:

$$[\eta] = \lim_{C \to 0} \frac{\ln z}{C} \approx 2.303\left(0.001K + 0.000075K^2\right)$$

Then the viscosity average molecular weight is calculated by Mark–Houwink equation:

$$\bar{M}_V = \left(\frac{[\eta]}{a}\right)^{\frac{1}{b}}$$

where $a = 1.4 \times 10^{-4}$ and $b = 0.7$, or $a = 3.1 \times 10^{-4}$ and $b = 0.61$ [23].

However, the weight average molecular weight is more frequently used in recent years. Table 2.13 lists the weight average molecular weights of PVP with different *K*-values.

2.8.1.2 Solubility

One of the outstanding properties of PVP is its universal solubility both in hydrophobic and in extremely hydrophilic solvents. Because the molecular structure of PVP has both hydrophilic imido groups and lipophilic alkyl groups, it can be dissolved in

Table 2.13: Weight average molecular weights of PVP with different *K*-values.

PVP grade	Weight average molecular weight
Povidone *K*-12	2,000–3,000
Povidone *K*-17	7,000–11,000
Povidone *K*-25	28,000–34,000
Povidone *K*-30	44,000–54,000
Povidone *K*-90	1,000,000–1,500,000

water as well as many organic solvents. PVP is miscible completely in water and some ordinary organic solvents such as ethanol and isopropanol. However, the solubility of PVP in acetone is as small as only 1–2%. The organic solvents that can dissolve more than 10% of PVP at room temperature are as follows:

Alcohols: methanol, ethanol, propanol, isopropanol, *n*-butanol, isobutanol, pentanol, BDO, cyclohexanol, ethylene glycol, propylene glycol, butylene glycol, glycerin

Carboxylic acids: formic acid, acetic acid, propionic acid, acrylic acid

Esters: ethyl lactate, vinyl acetate, γ-butyrolactone

Alcohol-ethers: glycol ether, diethylene glycol, triethylene glycol, polyethylene glycol 200, polyethylene glycol 400

Halogenated hydrocarbons: dichloromethane, chloroform, 1,2-dichloroethane

Amines: butylamine, cyclohexylamine, aniline, ethylenediamine, pyridine, morpholine, monoethanolamine, diethanolamine, triethanolamine, 2-amino-2-methylpropanol

Lactams: 2-pyrrolidone, NMP, *N*-vinylpyrrolidone

Nitrohydrocarbons: nitromethane, nitroethane

Amide: dimethylformamide, dimethyl acetamide

Sulfone: dimethyl sulfoxide

However, the solubility of PVP in the following solvents at room temperature is less than 1%:

Hydrocarbons: butane, pentane, hexane, cyclohexane, methylcyclohexane, heptane, benzene, toluene, xylene, solvent oil, kerosene, turpentine and liquid paraffin

Ethers: dimethyl ether, diethyl ether, tetrahydrofuran, ethyl vinyl ether, isobutyl vinyl ether, petroleum ether, dioxane

Halogenated hydrocarbons: carbon tetrachloride, chlorobenzene

Esters: methyl acetate, ethyl acetate, isobutyl acetate

In general, PVP is easier to dissolve in more polar solvents, but difficult to dissolve in weakly polar or nonpolar solvents. However, when a co-solvent is present, the solubility of PVP in weakly polar or nonpolar solvents can be increased.

To produce solutions of PVP from powder, it is advisable to add the powder to the solvent in portions while stirring, avoiding formation of lumps of powder. When using

PVP with higher molecular weight to make more concentrated solutions, the rate of dissolution can be significantly increased by elevation of the dissolution temperature to slightly above room temperature.

2.8.1.3 The distribution behavior of PVP in some two-phase systems

In the water–dichloromethane two-phase system, almost all PVP enters the water phase, which is the opposite of the monomer NVP. Almost all NVP enters the organic phase in this two-phase system. For the water–pentanol two-phase system, PVP also almost enters the water phase, while for the water–phenol two-phase system, PVP tends to enter the organic phase, and the partition coefficient reaches 10. When a large amount of acetone is added to the PVP aqueous solution, part of the PVP will precipitate out and form two phases. Using this property, PVP can be isolated from its aqueous solution. For a PVP aqueous solution with a certain molecular weight distribution, when acetone is added, the portion with the larger molecular weight is precipitated first. Therefore, fractional precipitation can be used to separate PVP with different molecular weights. Based on the same principle, acetone can be titrated into the PVP aqueous solution to determine the *K*-value and molecular weight of the PVP sample. The greater the amount of acetone consumed when turbidity occurs during titration, the smaller the *K*-value of PVP.

2.8.1.4 Surface activity

The molecular structure of PVP contains a strong polar acylamino group with a dipole distance of 40 C · m, which has the ability of hydrophilic and affinity polar groups, and its nonpolar alkyl group makes it lipophilic. The end of the lactam-based oxygen atom with the larger dipole moment is bare, while the end of the nitrogen atom is surrounded by methyl and methylene groups. The molecular structure of PVP makes it surface active. Although its ability to reduce surface tension is lower than that of low-molecular-weight surfactants, its solid surface adsorption ability and solid shielding ability due to its hydrophilic properties can make solid particles have excellent dispersion stability. As a polymer compound, it has the ability to widely adjust the rheological properties of dispersions or solutions. Its hydrogen bonding ability with many inorganic compounds and organic compounds provides excellent coagulation or solubilization effect, the former makes it clarify and stabilize in alcohol and beverages containing polyphenols, while the latter is widely applied in coprecipitation to prepare drug solid dispersions for improved solubility of poorly soluble drugs. Therefore, PVP has become one of the main varieties of polymer surfactants.

2.8.1.5 Complexation

The PVP molecule has a strong polar amide group and can bond with hydrogen, which allows it to have a strong complexing ability with many substances, especially compounds containing hydroxyl, carbonyl, carboxyl, amino and other active hydrogen atoms. PVP forms solid complexes with iodine, β-carotene, reserpine, tolbutamide, phenytoin, amorpholine, griseofulvin and various sulfonamide drugs. After complexation, the thermodynamic activity of these small molecular substances is reduced, and the stability as well as water solubility is improved. For example, iodine is easily sublimated, insoluble in water and easily soluble in chloroform, but the complex of PVP and iodine (PVP-I) is easily soluble in water, nonvolatile, and even unable to be extracted by chloroform. The complexation of PVP has been widely used in the field of medicine, and many insoluble drugs become soluble after complexing with PVP.

The complexes formed by PVP and polyphenols are usually insoluble in water. The stability of these complexes is related to the pH of the solution. PVP forms complexes insoluble in water and ethanol with polyacids such as polyacrylic acid, polymethacrylic acid and tannic acid in aqueous solution, but after neutralization with alkali, the complex reaction proceeds in reverse. These properties can be used to separate or remove these substances. Some dyes, especially direct dyes, acid dyes and sulfur dyes can also form complexes with PVP, so PVP can be used as a stripping agent and anti-staining agent.

Oxygen and nitrogen atoms in PVP molecules are typical coordination atoms, making them capable of forming complexes with many metals. Among them, the most studied are transition metals and precious metals. For example, PVP and Fe, Mn, Co, Ni and other transition metals form complexes. These complexes are prepared by the reaction of PVP with metal carbonyl compounds. The carbonyl group in the PVP molecule replaces the carbonyl group in the metal carbonyl compound. PVP metal complexes have important applications in the field of catalysis and have received increasing attention in recent years.

2.8.1.6 Chemical stability

Under normal circumstances, PVP solids are very stable, no change after heating at 100 °C for 16 h. If heated to 150 °C, or mixed with an initiator such as ammonium persulfate and heated at 90 °C for 30 min, a cross-linking reaction occurs and it becomes an insoluble cross-linked solid. In the presence of azo compounds, irradiation with ultraviolet light and gamma rays will produce a stable gel. Long-term grinding will lead to degradation of PVP, and the average molecular weight will decrease. For example, after grinding PVP *K*-90 in nitrogen for 100 h, the molecular weight will decrease from 1,000,000 to about 400,000, and after 200 h, it will drop to less than 100,000.

The aqueous solution of PVP is usually very stable, and there is no obvious change when heated to 100 °C, yet the color of the solution becomes light yellow after heating for a long time. Addition of some salts containing polyvalent anions to the PVP aqueous solution, such as sodium tripolyphosphate and sodium metasilicate, will cause precipitation.

In the presence of acids or bases, heating to 110 °C, part of the pyrrolidone rings in the PVP molecule are cleaved to form polymers of *N*-vinyl-γ-aminobutyric acid or salt. Under alkaline conditions, the following reaction occurs:

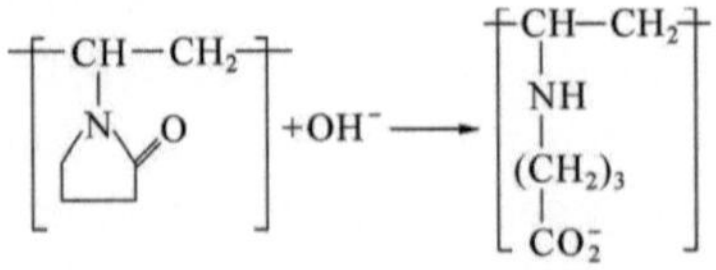

The conversion depends to the pH of the solution. For a 2% PVP aqueous solution, the conversion is 5% at a pH value less than 1, and the conversion is close to 20% at pH = 14.

For the aqueous solution of PVP with high molecular weight, the tendency of the viscosity to decrease slowly with time can be observed, indicating that the polymer is slowly degraded in the aqueous solution. A small degradation is also found in PVP solid powder with high molecular weight PVP after long-term storage.

2.8.1.7 Physiological inertness and safety

A large number of toxicological and biochemical studies shows the excellent physiological inertness and biocompatibility of PVP. The acute, subchronic and chronic toxicity of PVP orally or intraperitoneally is extremely low, with the only effect observed being diarrhea at high doses due to its osmotic action. LD_{50} values of PVP are in the range from 12 g kg^{-1} intraperitoneally in the mouse to 100 g kg^{-1} orally in the rat, suggesting it belongs to the actual nontoxic level according to the oral toxicity classification standard. PVP has no irritating or allergic effects on skin, mucous membranes and eyes in humans. Neither carcinogenic effect nor reproductive toxicity has been found following oral administration or intraperitoneal, subcutaneous and intravenous injections. Pharmacological studies have demonstrated that PVP has no effect on the central nervous system, respiratory system and blood circulation system.

PVP is not metabolized in the body. It is not absorbed by the gastrointestinal tract. No gastrointestinal toxicity has been reported. PVP with a lower molecular weight can be completely excreted in urine within a few days. PVP with a higher molecular weight will slowly accumulate in the body, mainly in reticuloendothelial tissue cells, especially temporarily in the spleen, liver, lymph nodes, bone marrow and kidneys. The degree of storage and disappearance is related to molecular weight, dosage and duration of intake. However, this accumulation did not cause damage or harmful effects on tissue morphology or function. Non-gastrointestinal administration studies have shown that PVP has no carcinogenic effect, and some experiments even observed that PVP has a certain tumor suppressing effect.

Because PVP and NVP copolymers are widely used as film-forming substances for hair sprays, the study of inhalation toxicity is of great significance. A large number of animal and clinical tests have shown that no inhalation of PVP caused pathological changes in the lungs. The US FDA announced that aerosol hairspray containing PVP as a film-forming agent does not pose a risk of damage to health under normal conditions of use.

The acceptable daily intake was adjusted to 0–50 mg kg^{-1} body weight by the FAO/WHO Joint Expert Committee on Food Additives (JECFA) in 1987.

In parenteral administration, PVP with low molecular weight is easily excreted from the renal system, while PVP with high molecular weight is excreted slowly. Therefore, it must be noted that when PVP is used as an injection to directly enter the human blood, its molecular weight is an important factor that must be considered. PVP does not maintain its original compact configuration like protein in aqueous solution, but exists in water like a polymer with flexible chains in a scattered spiral configuration, and its size depends on molecular weight. In saline, the sizes of PVP molecules with different molecular weights are between 1 and 100 nm. The inner diameter of the human kidney capillary is about 7 nm; that is, the molecular size of PVP entering the human blood circulation system should be less than 7 nm to be excreted smoothly through the kidney. The molecular weight corresponding to the 7 nm size is about 30,000; that is, the molecular weight of PVP as an injection component should be less than 30,000 [24]. In fact, polymer products are characterized by average molecular weight, and there is also the problem of molecular weight distribution. Therefore, to ensure that PVP products used for injections do not contain components with molecular weights greater than 30,000, the average molecular weight is required to be much less than 30,000.

2.8.2 Process for manufacture of PVP

The polymerization of NVP monomer has been initiated by many initiators including hydrogen peroxide–ammonia, organic peroxide, 2,2′-azobis(isobutyronitrile), persulfates, sodium peroxide, and sodium sulfite. PVP series products with different molecular weights can be obtained by controlling different polymerization process conditions. In order to obtain better quality polymer products, the purity of NVP monomer is very important. In general, the purity of NVP monomer should be more than 99.8%, and NVP monomer should not be stored for too long to avoid a decrease in polymerization activity. Before the polymerization reaction of NVP monomer, vacuum distillation is often required in advance. The first purpose is to further purify the monomer. The second purpose is to remove the polymerization inhibitor added to the NVP monomer for preventing polymerization during storage and transportation.

2.8.2.1 The homopolymerization process of *N*-vinylpyrrolidone

NVP monomer is extremely prone to polymerization. Heating NVP monomer or adding an initiator to NVP monomer can initiate the homopolymerization of NVP. The homopolymerization is a polymerization of NVP monomer itself. The resulting polymers are PVP with different average molecular weights, that is, PVP with different *K*-values.

Similar to other polymer synthesis methods, NVP homopolymerization methods also include bulk polymerization, solution polymerization and suspension polymerization.

2.8.2.1.1 Bulk polymerization

Heating pure NVP monomer above 140 °C can initiate the polymerization reaction by heat. When NVP monomer and peroxide are heated together to 110 °C, bulk polymerization also occurs. The reaction is strongly exothermic, and the reaction temperature can rise to 190 °C by itself, leading a polymer in the molten state. After cooling, it is crushed to obtain a powdery polymer product. In the bulk polymerization, the heat transfer is difficult and local overheating occurs because of high viscosity of the reaction system, resulting in a yellow polymer with low *K*-value and high residual monomer content. The polymer with poor quality cannot meet the requirements of the market; therefore, bulk polymerization of NVP has no practical application value in industrial production.

2.8.2.1.2 Solution polymerization

NVP monomer can be polymerized in many solvents, such as water, ethanol and isopropanol. Since water is the cheapest and safest solvent, it is most common to carry out the polymerization in aqueous solution. The main factors affecting the polymerization of NVP solution include the solvent, initiator and dosage, monomer concentration, the pH value of the reaction system, the polymerization temperature and time.

The influence of the solvent on the NVP polymerization reaction is mainly manifested in the influence of different kinds of solvents on the molecular weight of the PVP produced. This is because the polymerization mechanisms of NVP are different when different solvents are used, which lead to different activation energies and rates of radical chain transfer and chain termination reactions, so the molecular weights of the polymers obtained are also different. The chain growth and chain termination activation energies of NVP solution polymerization in different solvent systems are shown in Table 2.14.

Table 2.14: Chain growth and chain termination activation energy of NVP solution polymerization (NVP concentration of 50%).

Solvent	Activation energy /kJ · mol^{-1}	
	Chain termination	Chain growth
Water	19	68
Isopropanol	13	40 ± 2
Methanol	9	32 ± 3
Ethyl actate	8	23

Fikentscher first developed a process for polymerization of NVP in aqueous solution using an initiator system of hydrogen peroxide and ammonia. The polymerization mechanism is as follows:

Initiation:

$$H_2O_2 \longrightarrow HO\cdot + \cdot OH$$

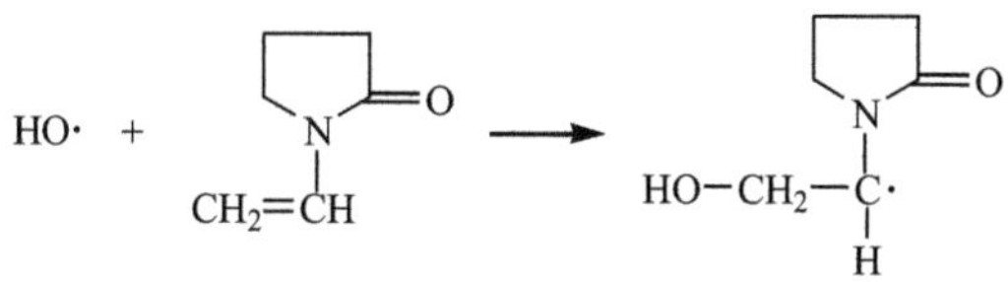

Propagation:

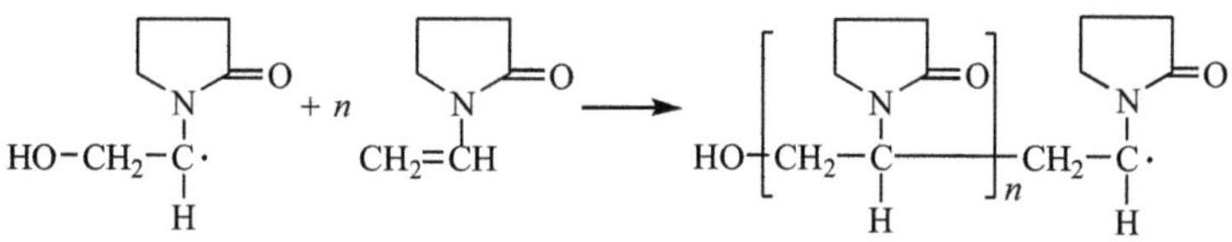

Termination:

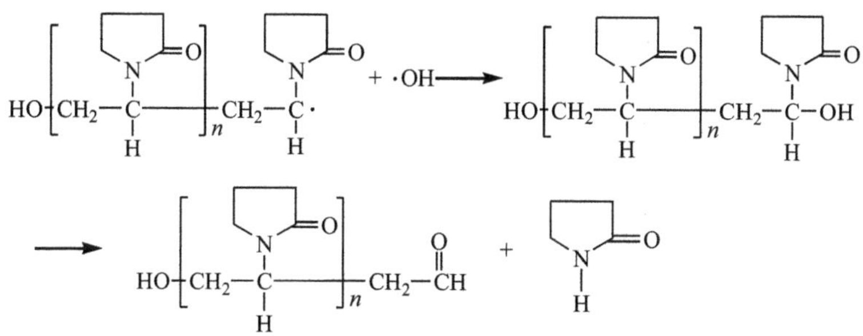

The hydroxyl groups produced by the decomposition of hydrogen peroxide are called end groups in the polymer product. Since the pyrrolidone ring is cleaved during the chain termination reaction, the second terminal group of PVP obtained by polymerization of NVP in aqueous solution initiated by hydrogen peroxide is an aldehyde group, and the polymer product also contains a small amount of 2-pyrrolidone.

When water is used as a solvent and azo compounds (azobisisobutyronitrile, azobisisoheptanonitrile, etc.) are used as initiators, the polymerization mechanism is as follows:

Initiation:

$$NR{=}NR \xrightarrow{\text{Heat}} 2R\cdot + N_2\uparrow$$

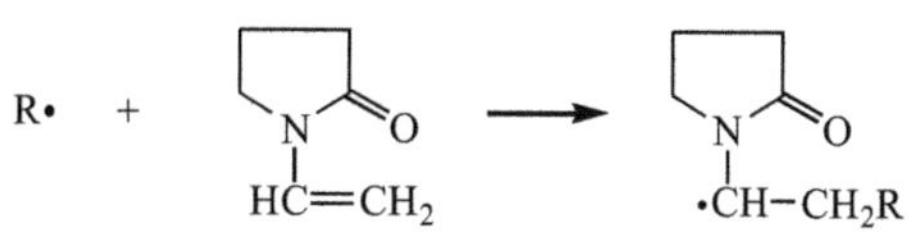

Propagation:

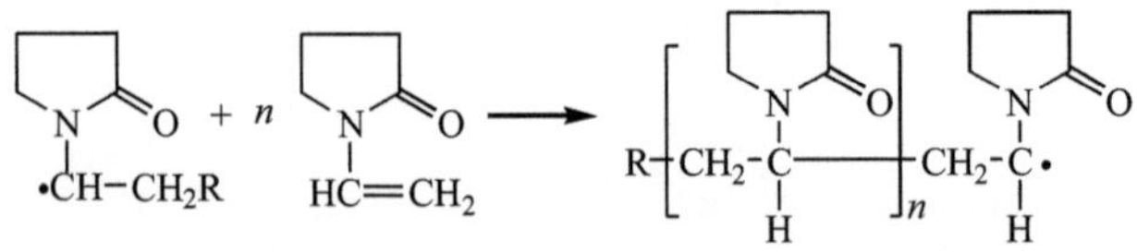

Termination:

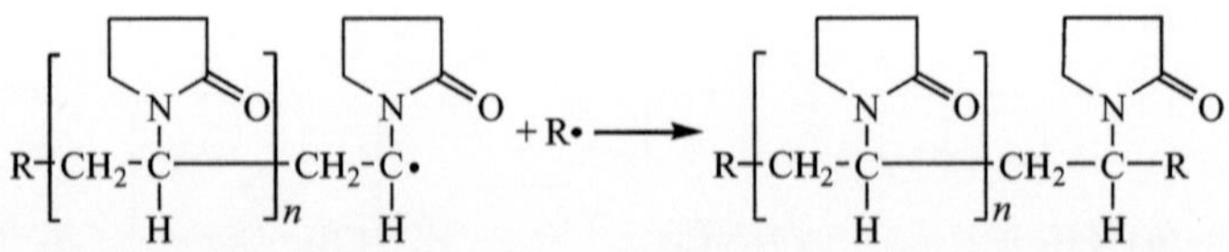

Using some organic peroxides as initiators, PVP with low molecular weight (*K*-value of 10–30) can be obtained by polymerization of NVP in organic solvents. Various organic peroxides initiators including *tert*-butyl hydroperoxide, cumyl hydroperoxide, di-*tert*-butyl hydroperoxide, dibenzoyl peroxide, *tert*-butyl peroxypivalate are used. The polymerization mechanism in organic solvents is as follows:

Initiation:

$$ROOR \xrightarrow{\text{Heat}} RO\cdot + \cdot OR$$

$$RO\cdot + SH \longrightarrow ROH + S\cdot$$

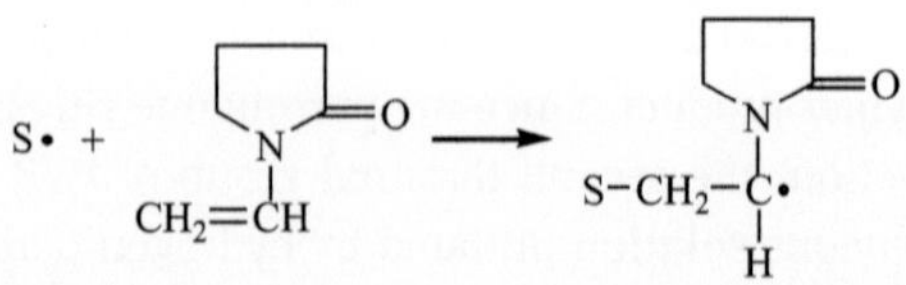

Propagation:

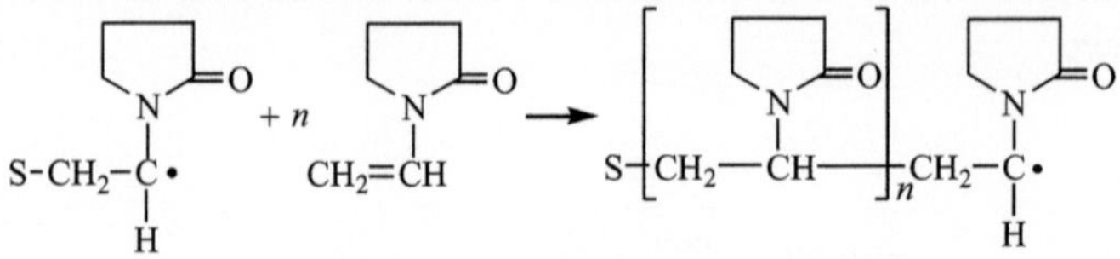

Termination:

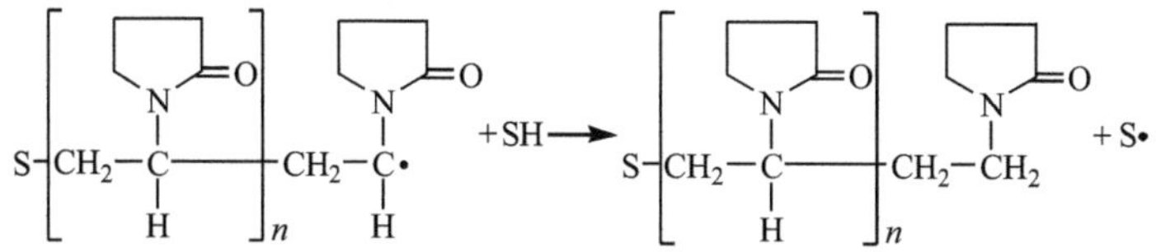

In the organic solvent SH, the organic peroxide generates an alkoxy group upon heating, and the alkoxy group reacts with the solvent to generate a solvent radical S·. The initiation of polymerization is carried out by solvent radicals rather than by alkoxy groups. The termination of the chain is accomplished by accepting a proton in the solvent molecule, resulting in a polymer molecule containing alkyl end groups and a solvent radical. The solvent radical continues to initiate polymerization. Since there is no cleavage of the pyrrolidone ring, the resultant polymer is relatively pure, which is absence of 2-pyrrolidone and aldehyde end groups. Therefore, the polymer is more stable under oxidizing conditions.

In general, the industrial production of PVP is by solution polymerization [25]. The trace amount of residual monomer can be removed by adsorption [26, 27].

2.8.2.1.3 Suspension polymerization

In a suspension polymerization process, heptane is used as the continuous phase, and NVP is dispersed in heptane. In order to achieve a good dispersion effect, a copolymer of vinylpyrrolidone and long-chain olefins is also added as a dispersant. PVP with a *K*-value of 90 can be prepared by suspension polymerization of NVP in the presence of azobisisobutyronitrile initiator. A small amount of water is also added to the suspension for adjusting the molecule weight of PVP. Since the PVP produced by suspension polymerization is insoluble in heptane, it can be directly filtered and dried to obtain PVP powder. Examples of suspension polymerization are as follows:

To a straight four-necked flask equipped with stirring, thermometer and nitrogen inlet, predetermined amounts of heptane, dispersant Antaron V-216, NVP monomer and pure water was added. Nitrogen was introduced into the flask to remove oxygen in the system. A constant temperature water bath was applied to heat the suspension. when the temperature was elevated to 75 °C, azobisisobutyronitrile initiator was added, and the polymerization reaction was carried out at this temperature for 8 h. After reaction, the reaction product was filtered, washed and dried at a temperature of 50 °C under vacuum to obtain PVP *K*-90 solid powder.

The main factors affecting the molecular weight or *K*-value of PVP are the dosage of initiator and the amount of water added. The less the initiator is added, the greater the *K*-value of the product PVP, that is, the greater the average molecular weight of the polymer (Table 2.15). With an increase of the amount of water added, the *K*-value of the resultant PVP is increased, and the polymer with greater average molecular weight is obtained (Table 2.16).

Table 2.15: The influence of initiator dosage on *K*-value of PVP.

No.	AIBN/%	*K*-value
1	0.12	84.9
2	0.19	82.9
3	0.26	73.6

Table 2.16: The influence of water addition on *K*-value.

No.	Water/%	*K*-value
4	4.7	83.6
5	6.8	86.2
6	8.9	88.7
7	10.9	90.1
8	14.6	92.7
9	16.4	93.9

2.8.2.2 Copolymerization

NVP are easily copolymerized with other monomers to form polymers containing both NVP structural units and other comonomer structural units. NVP can be copolymerized with only one other monomer to obtain a binary copolymer, or it can be copolymerized with two other monomers to obtain a terpolymer. In general, the properties of a copolymer are between those of homopolymers of each comonomer. By copolymerization, polymer properties can be modified to better meet the actual application requirements of the product and the application range of the polymer is further expanded. Many monomers that can be copolymerized with NVP, which are mainly as follows:

(1) Organic acids: acrylic acid, methacrylic acid, crotonic acid, etc.
(2) Esters: vinyl acetate, β-hydroxyethyl methacrylate, methyl methacrylate, vinyl propionate, dimethylaminoethyl propylene ester and so on.
(3) Alcohols and ethers: propenyl alcohol, vinyl phenyl ether, vinyl isobutyl ether, vinyl cyclohexyl ether and so on.
(4) Amides: acrylamide and its derivatives, vinyl benzimide, vinyl caprolactam, methylene bisacrylamide and so on.
(5) Unsaturated hydrocarbons: styrene, hexadecene, divinylbenzene, and so on.
(6) Others: acrylonitrile, maleic anhydride, vinyl imidazole, vinyl trimethyl silane, and so on.

The copolymerization reaction can be carried out in a solution, emulsion or suspension. Commonly used solvents are water, methanol, ethanol, isopropanol, benzene, toluene, acetone, ethyl acetate and so on. The free radical initiator used in the homopolymerization of NVP can also be used in the copolymerization reaction, but the commonly used initiator is azobisisobutyronitrile because of its low price and easy availability, fast reaction speed, high conversion and small dosage.

In the copolymerization reaction, if the ratio of the comonomer is different, and the composition, performance, and use of the obtained copolymer are also different. Therefore, the synthesis of NVP copolymers needs to control certain copolymerization reaction conditions to keep the composition of the copolymer within a certain range to ensure the uniformity of the copolymerized product. For this reason, it is necessary to select a suitable molar ratio of the comonomers, and match the appropriate copolymerization reaction conditions to obtain an ideal product according to the difference of the reactivity ratio between the comonomers. Usually, two methods can be applied to control the composition of the copolymer and keep it constant. One method is to control the conversion, i.e., the polymerization is carried out at a suitable monomer ratio to a certain reaction conversion, and then it is terminated by addition of a polymerization inhibitor or a chain transfer agent. This method is suitable for systems where the relationship between the composition of the copolymer and the conversion is relatively flat. The other method is to add monomers in batches, which means that the monomer with faster reaction (higher reactivity ratio) is added to the monomer with slower reaction (lower reactivity ratio) in batches or continuously in a dropwise manner for copolymerization. This method is suitable for systems where the slope of the relationship curve between copolymer composition and conversion is relatively large. The reactivity ratio of NVP and some monomers is shown in Table 2.17.

Table 2.17: Reactivity ratio when NVP (M1) is copolymerized with other monomers.

Comonomer (M2)	r_1	r_2
Acrylonitrile	0.06	0.18
Allyl alcohol	1.0	0.00
Vinyl acetate	2.0	0.24
Propylene acetate	1.6	0.17
Methyl methacrylate	0.005	4.7
Propylene diacetate	0.92	0.94
Vinyl laurate	1.15–1.3	0.01
Maleic anhydride	0.16	0.08
Styrene	0.045	15.7
Trichloroethylene	0.54	<0.01
Vinyl chloride	0.38	0.53
Butyl vinyl ether	3.30	0.205
Cyclohexyl vinyl ether	3.84	0.00
Isopropyl vinyl ether	1.68	0.00
Phenyl vinyl ether	4.43	0.22
N-Vinyl benzimide	0.35	0.04

Graft copolymerization of NVP can be carried out by radiation. NVP can also be directly grafted by free radical polymerization in the presence of azo initiators or peroxide initiators. The polymers usually grafted by NVP are polyacrylonitrile, nylon, cellulose, gelatin and so on. NVP grafting can improve the hygroscopicity and hand feeling of some polymers.

Although many NVP copolymers can be obtained by free radical polymerization, only a few of them are produced in industry. For example, the copolymer of NVP and vinyl acetate, usually referred to as PVP/VA or VAP resin, changes the hygroscopicity and solubility of the homopolymer. By changing the ratio of NVP to vinyl acetate, VAP resins with different properties and uses can be manufactured, which are widely used in cosmetics such as hair care and skin care. The copolymerization of *N*-vinylpyrrolidone and vinyl acetate is usually carried out in a solvent such as ethanol or isopropanol in the presence of azobisisobutyronitrile or *tert*-butylperoxypivalate initiators.

The terpolymer of NVP, β-hydroxyethyl methacrylate and styrene has become an important soft material for artificial cornea and contact lenses due to the characteristics of water absorption, water retention, oxygen permeability, biocompatibility, non-irritation and good strength.

Other copolymers produced in industry include NVP/cationic monomer copolymer (Figure 2.12) and NVP/methacrylamide/vinyl imidazole copolymer (Figure 2.13). They have better curl retentions for hair.

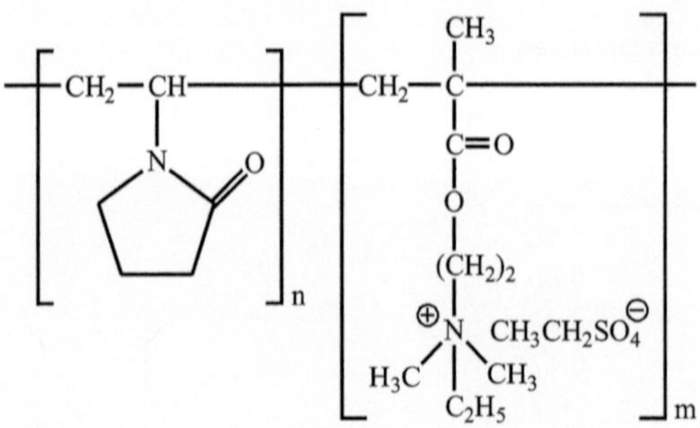

Figure 2.12: Copolymer of NVP and cationic monomer.

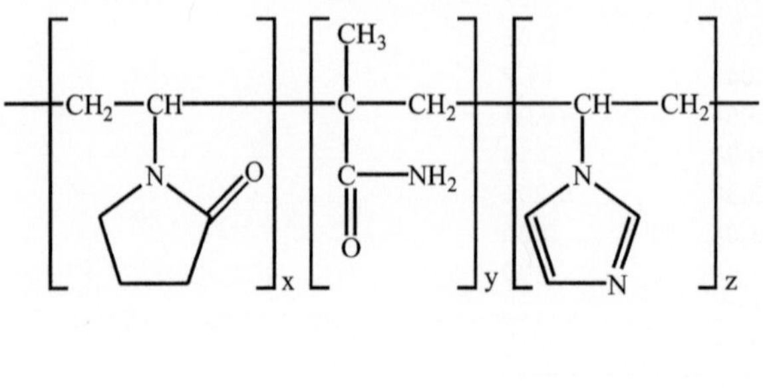

Figure 2.13: Copolymer of NVP, methacrylamide and vinyl imidazole.

2.8.2.3 Cross-linking polymerization of NVP

The cross-linked polymer of NVP is known as polyvinylpolypyrrolidone (PVPP), insoluble polyvinylpyrrolidone (insoluble PVP), cross-linked PVP, or crospovidone. The cross-linking polymerization of NVP can be carried out in two ways. One is addition of a bifunctional cross-linking agent to the NVP aqueous solution, and cross-linking polymerization is initiated by azobisisobutyronitrile under the protection of nitrogen. Commonly used crosslinking agents include divinylbenzene and *N,N′*-methylenebisacrylamide. In this way, a gel-like polymer is usually obtained with low crosslink density. High-density cross-linking is achieved by the second method, namely popcorn polymerization or proliferous polymerization.

Popcorn polymerization does not require the addition of cross-linking agent and initiator. NVP itself generates cross-linking agents during the process. In the presence of alkaline hydroxide, heating the NVP aqueous solution to above 100 °C can produce small amounts of bifunctional monomers 1-vinyl-3-ethylenepyrrolidone and ethylene-bis-3-(*N*-vinylpyrrolidone), as shown in Figure 2.14. The contents of these bifunctional compounds in the reaction mixture are approximately 1.5% and 0.1%, respectively. They act as cross-linking agents and undergo cross-linking polymerization with NVP monomer to produce PVPP, a polymer with a high degree of self-cross-linking. The polymer is insoluble in water and organic solvents, and has been widely used as a clarifying and purifying agent for beer and beverages or disintegrant for medicines. It is also possible to add a small amount of [<2% (mol)] functional group crosslinking agent such as *N,N′*-methylenebisacrylamide to the NVP aqueous solution for crosslinking polymerization in the absence of a initiator. After heating,

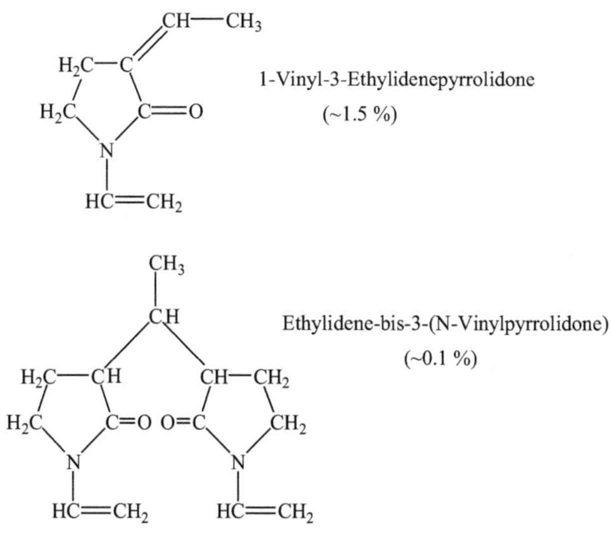

Figure 2.14: Two bifunctional monomers produced by NVP itself.

crosslinking polymerization occurs, and the crosslinking density of the resulting polymer is much higher than that by azobisisobutyronitrile initiator.

2.8.3 Uses of polyvinylpyrrolidone series products

PVP has many excellent properties, making it widely used in many fields [28–37].

2.8.3.1 Pharmaceutical industry, medical and health fields

PVP has good biocompatibility and stable chemical properties. It is widely used as a binder, thickener, solubilizer, dispersant, stabilizer, film-forming agent and detoxifier for eye drops in in the pharmaceutical industry.

PVPP is used as a disintegrant for drugs. It is also used for stabilization of vitamins. Moreover, micronized PVPP is used as an active substance for treatment of diarrhea.

PVP-I is a nontoxic, nonirritating and highly effective disinfectant that is currently popular in the world. It is used for the disinfection of the skin of the surgical site, the disinfection of medical equipment, public health and the disinfection of the pharmaceutical and food industries. PVP-I can also be used as an external medicine. Three dosage forms including liquid, paste and tablet are available, which are used for external antibacterial and anti-inflammatory, treatment of burns and vaginitis. The currently used drug dosage forms are shown in Table 2.18.

Table 2.18: Application of PVP in medicine.

Dosage form	PVP dosage/%	PVP ingredient	Function
Tablets	0.5–5	PVP *K*-30, *K*-90	Bonding, shaping and solubilization
Granules	0.5–5	PVP *K*-30, *K*-90	Bonding, shaping and solubilization
Syrup, oral solutions	2–10	PVP *K*-30 PVP *K*-90	Dispersion, uniformity, stability, thickening and taste masking
Injection	5–15	PVP *K*-12, *K*-15, *K*-17	Solubilizing and dispersing
Capsules	1–2	PVP *K*-30	Bonding, improvement of flowability
Ophthalmic medicine	0.05–12	PVP *K*-30, *K*-60	Reduction of irritation, improvement of stability
Disinfectant	–	PVP-I	Reduction of toxicity and irritation, improvement of stability
Film forming material	0.5–5	PVP, PVP/VA	Improvement of the adhesion of the film to the surface, improvement of stability and film uniformity
Disintegrant	1–5	PVPP	Swelling

Table 2.18 (continued)

Dosage form	PVP dosage/%	PVP ingredient	Function
Vitamin formulations	–	PVP *K*-90, PVPP	Stabilization
Coprecipitation agent	–	PVP *K*-25, *K*-30	Improvement of the dissolution and bioavailability of insoluble drugs

2.8.3.2 Cosmetics

PVP contributes to the softness and lubrication of the skin, promotes the natural humidity balance in skin cosmetics, and reduces the irritation of cosmetics to the skin. It is a cosmetic antidote with superior performance. PVP is used in shampoo to improve the stability of foam, increase the gloss and smoothness of hair and make hair elegant. PVP-I can be used as an anti-itch and anti-dandruff agent. PVP and its copolymers are used for hair styling and hair care. The film formed on the hair is elastic and shiny, and has excellent combing performance. Therefore, they are indispensable raw materials for hair cosmetics such as styling hair cream, hairspray and mousse.

2.8.3.3 Foods and beverages

PVPP can be used as a clarifying, purifying agent and stabilizer for beer, wine and fruit juice. The PVPP-treated beer has a long storage time (more than 12 months) and has obvious improvement of transparency, stability and flavor. Currently, more than 38 countries in the world have successively adopted the PVPP adsorption filtration process for beer stabilization.

2.8.3.4 Textile, printing and dyeing industry

The introduction of PVP into synthetic fibers can increase the dye susceptibility and antistatic properties of synthetic fibers, and improve the hygroscopicity, wrinkle resistance, styling and easy washing of synthetic fibers. For example, polyamide fibers and polyester fibers are popular for their high strength, high fineness, abrasion resistance, thermoplasticity, good gloss and bright dyeing, but their moisture absorption is poor. If they are used as clothing, a layer of water vapor from the air is often attached to the surface of clothing, and the clothing with poor hygroscopicity is easy to get wet and cling to the skin especially in hot and humid weather, leading to a discomfort. One way to solve this problem is to add 4–8% of PVP *K*-60 before fiber spinning, so that synthetic fibers such as polyamide fiber and polyester fiber have high moisture absorption and washing resistance. The graft modification of nylon 66 with NVP not only improves dyeability, but also improves moisture absorption, wrinkle resistance, styling, easy washing, bleaching strength and solar resistance.

PVP has a strong affinity with many organic dyes, which is mainly attributed to the binding force between the lactam structure in the PVP molecule and the hydroxyl, amino and carboxyl groups in the dye molecule. This property of PVP can be used to improve the dyeability and hydrophilicity of many hydrophobic synthetic fibers, so that synthetic fibers can be dyed evenly, and the depth of dyeing can be increased. For example, for polypropylene fibers that are difficult to dye, adding 7.5–10% of NVP copolymer to the resin can increase the dyeing depth of the fiber with acid dyes by nearly 10 times and the dyeing depth with disperse dyes by 3–3.7 times.

PVP is also used as a post-treatment agent for fabrics. Generally, in fabric finishing, cationic finishing auxiliaries and anionic fluorescent brighteners are incompatible with each other, and fabrics cannot be processed in the same bath. In the presence of 0.1–2% of PVP, various additives such as softener, antistatic agent, resin finishing agent and fluorescent brightener can be processed and finished in one bath at the same time. The treated fabric has excellent properties such as whiteness, fluorescence, softness, strength and antistatic properties.

2.8.3.5 Adhesives

PVP is widely used in various adhesive formulations. Its application can be divided into two categories: one is to use PVP as the main component of the adhesive, and the other is to use the bonding properties of PVP in other products that require adhesive components, such as coatings, inks, various tablets, granules and sintered materials.

As the main component of adhesives, PVP is most widely used in industry including solid glues, pressure-sensitive adhesives and rewetting adhesives. Solid glue is a new type of office glue, and it has been widely popular in the world, especially suitable for children. To compared with liquid glue, solid glue is easy to carry and use, and suitable for all kinds of paper. It has strong initial adhesion, and the paper is smooth and wrinkle-free after bonding. Although solid glues based on polyvinyl alcohol or other raw materials have appeared on the market, their performance is far inferior to PVP solid glues. Another adhesive with polymers of NVP as the main component is pressure-sensitive adhesive. Addition of NVP polymers to the pressure-sensitive adhesive formulation can improve the initial adhesion and strength of the tape. Furthermore, polymers of NVP are also used as the main components in rewetting adhesives.

2.8.3.6 Pigments and coatings

In the preparation of organic pigments, PVP is often used as a surface coating agent and dispersant for pigments. The surface-treated pigment has excellent tinting power and gloss. For example, 0.2% of PVP *K*-30 was added to a lake blue pigment, which increased the coloring power by 20%. PVP has excellent dispersing effects on carbon black, phthalocyanine pigments, titanium dioxide, etc. After addition of PVP, carbon black is not easy to gelatinize during ball milling. The resultant carbon black has a smaller particle size and better stability during storage, avoiding the problems of easy flocculation, precipitation and agglomeration of carbon black suspension. The pigment dispersion made by kneading and pulverizing phthalocyanine pigment or titanium dioxide, cellulose acetobutyrate and PVP has good stability, which is used for the coloring of polyurethane solutions.

In the field of coatings, PVP can prevent gelation and act as a protective colloid in many formulations.

2.8.3.7 Detergent

PVP can be used as a detergent builder. It can prevent dirt from re-deposition, and the effect is much better than the commonly used carboxymethyl cellulose. It can effectively prevent the occurrence of color conversion or white background contamination in the washing after printing and dyeing. PVP also helps the brighteners in the detergent to adhere to the fibers, and reduce the irritation of the detergent to human skin.

2.8.3.8 Use in production of polymers

In the polymerization process, PVP can be used as a viscosity regulator, dispersion stabilizer and particle size regulator. The resin produced by this method has improved strength, transparency and color sensitivity. In general, PVP with a larger molecular weight is used as a protective agent and stabilizer for preventing the precipitation of the suspension or the aggregation of larger particles in the emulsion and dispersion, while PVP with a low molecular weight is more suitable for use as a dispersant, especially for the dispersion of low-density substances.

2.8.3.9 Agriculture

PVP is nontoxic to plants and can be used in agricultural sprays, compound fertilizers and wettable formulations for dusting and seeding. PVP film can prevent the leaves from withering during plant transplantation and protect the growth of plants. Applying it on the seeds can protect biological functions and reduce damage during soaking.

PVP-I is applied as high-efficiency fungicides and insecticides to kill nematodes, larvae of cereal worms and certain insects, and can control the pathological conditions that produce fungi in plants. In aquaculture, PVP-I can be used for disinfection, which greatly reduces the diseases caused by bacteria on fish, bullfrogs, prawns and other aquatic products.

2.8.3.10 New material field

PVP can be used for separation membranes, medical polymer materials and the blending of polymers. At present, the applications of membrane separation technology in the pharmaceutical industry, sewage treatment and water purification have been very common. Among them, nanofiltration and ultrafiltration membranes are important components in the membrane separation industry. PVP is used as its film-forming porogen, and its consumption has now exceeded 3,000 t a^{-1}. In addition, PVP is also used in the preparation of membrane materials for kidney dialysis.

The copolymer of NVP and methacrylic acid hydroxymethyl ester or glycol ester can be made into super absorbent soft contact lens with a water absorption percentage of 90%, having good water permeability, air permeability and mechanical strength.

PVP is widely used as a dispersant and stabilizer in the preparation of nano materials. It is used as a dispersant for cathode materials of lithium batteries in the new energy industry.

PVP and active metal complexes used as catalysts are also attracting attentions [32–36]. It is also a promising adsorption material for waste water treatment [37, 38].

2.8.3.11 Other uses

PVP is a main polymer quenching agent. PVP aqueous solution used as a quenching agent was first applied in the United States. This type of quenching agent has many advantages such as safe operation, large flexibility, no pollution and low cost.

PVP is also used as paper auxiliary in paper industry.

2.8.4 PVP production status and market

Currently, PVP is mainly produced by Ashland (former ISP) and BASF. The annual production output of PVP series is more than 20,000 tons for each company. In Japan, the production of PVP started in 2003, and the capacity was said to be expanded to 3,500 t a^{-1} in 2007. The product was mainly PVP *K*-90 with high molecular weight. Russia and Iran also have small production facilities. Russia mainly produces PVP *K*-15 and *K*-17 with small molecular weights, having a production capacity of about 800 t a^{-1}. India has a production capacity of 1,200 t a^{-1}, and its production technology was purchased from Xiamen University, China.

The largest consumption of PVP in the United States and European Union is the daily chemical industry, accounting for more than 50% of the total, followed by the pharmaceutical industry, accounting for 20–30%, and the rest is used in beverages, papermaking, textiles and others.

In China, the development, production and application of PVP started late. The development of NVP and PVP began in the mid-1990s. The early production of NVP monomer was mainly by amination of γ-butyrolactone with ethanolamine, but later it was eliminated because of its high cost. In 2003, CNSG Anhui Hong Sifang Co., Ltd. adopted the technology developed by Xiamen University to launch the NVP and PVP production installations by vinylation of 2-pyrrolidone with acetylene and successfully put them into production. The total production output of PVP was only 3,200 tons in

2006. It increased to 5,000 tons in 2008, 8,050 tons in 2014, and more than 11,000 tons in 2018. At present, NVP and PVP manufacturers include NKY Medical Group, CNSG Anhui Hong Sifang, Jiaozuo Zhongwei, Jiaozuo Yuanhai, Zhangzhou Huafu, Quzhou Jianhua Nanhang, Sichuan Tianhua, and Wulanchabu Kema. In addition, there are also some companies that purchase NVP monomer from other manufacturers for polymerization to produce PVP.

References

[1] Zhang L. Analysis on the development advantages of 1,4-butanediol, a downstream product of acetylene, China Chlor-Alkali 2011, 5, 19–21.

[2] Bai G. 1,4-Butanediol, Tetrahydrofuran and its Industrial Derivatives, Chemical Industry Press, Beijing, China, 2012.

[3] Li Y, Wu X. Analysis of the supply and demand of 1,4-butanediol at home and abroad, Acetaldehyde Acetic Acid Chemical Industry 2020, 2, 6–11.

[4] Hunter S E, Ehrenberger C E, Savage P E. Kinetics and mechanism of tetrahydrofuran synthesis via 1,4-butanediol dehydration in high temperature water, Journal of Organic Chemistry 2006, 71, 6229–6239.

[5] Vaidya S H, Bhandari R V, Chaudhari R V. Reaction kinetics studies on catalytic dehydration of 1,4-butanediol using cation exchange resin, Applied Catalysis A 2003, 242, 321–328.

[6] Limbeck U, Altwicker C, Kunz U, Holmann U. Rate expression for THF synthesis on acidic ion exchange resin, Chemical Engineering Science 2001, 56, 2171–2178.

[7] Kudchadker S A. Thermodynamic properties of oxygen compounds. II. beta-Propiolactone, gamma-butyrolactone, and gamma-butyrolactam (2-pyrrolidone), Thermochim. Acta 1975, 12, 11–17.

[8] Lebedev B V, Yevstropov A A. Thermodynamics of β-propiolactone, γ-butyrolactone, δ-valerolactone, and ε-caprolactone from 13.8 to 340 K, Journal of Chemical Thermodynamics 1983, 15, 115–117.

[9] Ichikawa N, Sato S, Takahashi R, Sodesawa T, Inui K. Dehydrogenative cyclization of 1,4-butanediol over copper-based catalyst, Journal of Molecular Catalysis. A, Chemical 2004, 212(1–2), 197–203.

[10] Liu K-C, Taylor P D. Process for the preparation of 2-pyrrolidone. US 4,824,967, 1989.

[11] Rudloff M, Stops P, Henkes E, et al. Method for the production of 2-pyrrolidone. US 7,164,031, 2007.

[12] Ou Y, Zhu Y, Li C, Wang X. Effect of water on hydrolysis of N-methylpyrrolidone, Journal of Lanzhou University of Technology 2018, 44(2), 70–73.

[13] Bergfeld M Process for producing pyrrolidone and N-alky pyrrolidones. US 5478950. 1995.

[14] Yoon Y-S, Shin H K, Kwak B-S. Ring conversion of γ-butyrolactone into N-methyl-2-pyrrolidone over modified zeolites, Catalysis Communications 2002, 3(8), 349–355.

[15] Fandary M S, Al-Jimaz A S, Al-Kandary J A, Fahim M A. Extraction of pentylbenzene from high molar mass alkanes (C14 and C17) by N-methyl-2-pyrrolidone, The Journal of Chemical Thermodynamics 2006, 38(4), 455–460.

[16] Tan L S, Shariff A M, Lau K K, Bustam M A. Impact of high pressure on high concentration carbon dioxide capture from natural gas by monoethanolamine/N-methyl-2-pyrrolidone solvent in absorption packed column, International Journal of Greenhouse Gas Control 2015, 34, 25–30.

[17] Liu M, Liu J, Gai Y. Advances in development and petrochemical industrial applications of N-methyl pyrrolidone, Applied Chemical Industry 2017, 46(5), 991–994.

[18] Ao S, Li S. Measurement and correlation of solid–liquid phase equilibria for binary and ternary systems consisting of N-vinylpyrrolidone, 2-pyrrolidone and water, Chinese, Journal of Chemical Engineering 2018, 26, 806–811.
[19] Li S Preparation method of N-vinylpyrrolidone. CN 200810072060.9, 2012.
[20] Li S Separation and purification method of N-vinylpyrrolidone. CN 200810072061.3, 2010.
[21] Su A, Li S. The process of purification of N-vinylpyrrolidone by crystallization, Chemical Industry and Engineering Progress 2018, 37(2), 492–496.
[22] Ledovskaya M S, Voronin V V, Rodygin K S. Methods for the synthesis of O-, S- and N-vinyl derivatives, Russian Chemical Reviews 2018, 87(2), 167–191.
[23] Robinson B V, Sullivan F M, Borzelleca S S L. PVP – A Critical Review of the Kinetics and Toxicology of Polyvinylpyrrolidone (povidone), Lewis Publishers, Chelsea, Michigan, USA, 1990.
[24] Haaf F, Sanner A, Straub F. Polymers of N-vinylpyrrolidone: synthesis, characterization and uses, Polymer Journal 1985, 17(1), 143–152.
[25] Li S, Shi C, Lin H. Research on the preparation of polyvinylpyrrolidone, Chemistry World 1999, 40(4), 201–204.
[26] Li S, Umereweneza D. Adsorption of N-vinylpyrrolidone from polyvinylpyrrolidone solution onto bamboo-based activated carbon, Separation Science and Technology 2012, 47, 104–111.
[27] Li S Removal method of residual monomers in vinyl pyrrolidone polymer. CN 200910111011.6, 2010.
[28] Burnett C L. PVP (polyvinylpyrrolidone), International Journal of Toxicology 2017, 36(S2), 50S-51S.
[29] Cui Y, Yi G, Liao L. Synthesis and Application of Polyvinylpyrrolidone, Science Press, Beijing, China, 2001.
[30] Han Y, Song T, Yao X. Application of polyvinylpyrrolidone in traditional Chinese medicine preparations, Journal of Traditional Chinese Medicine 2008, (04), 58–60.
[31] Lu W. Application of polyvinylpyrrolidone K30 in the preparation of metronidazole tablets, Zhongnan Pharmaceutical 2005, (06), 354–355.
[32] Hu J-C, Cao Y, Yang P, et al. A novel homogeneous catalyst made of poly(N-vinyl-2-pyrrolidone)-$CuCl_2$ complex for the oxidative carbonylation of methanol to dimethyl carbonate, Journal of Molecular Catalysis. A, Chemical 2002, 185(1–2), 1–9.
[33] Evangelisti C, Panziera N, D'Alessio A, et al. New monodispersed palladium nanoparticles stabilized by poly-(N-vinyl-2-pyrrolidone): preparation, structural study and catalytic properties, Journal of Catalysis 2010, 272, 246–252.
[34] Kang S K, Kang Y S. Silver nanoparticles stabilized by crosslinked poly(vinyl pyrrolidone) and its application for facilitated olefin transport, Journal of Colloid and Interface Science 2011, 353, 83–86.
[35] Rakap M. Poly(N-vinyl-2-pyrrolidone)-stabilized palladium–platinum nanoparticles-catalyzed hydrolysis of ammonia borane for hydrogen generation, Journal of Power Sources 2015, 276(2), 320–327.
[36] Su X, Li S. PVP-stabilized Co-Ni nanoparticles as magnetically recyclable catalysts for hydrogen production from methanolysis of ammonia borane. International Journal of Hydrogen Energy 2021, 46, 14384–14394.
[37] Essawy H A, Ibrahim H S. Synthesis and characterization of poly(vinylpyrrolidone-co-methylacrylate) hydrogel for removal and recovery of heavy metal ions from wastewater, Reactive & Functional Polymers 2004, 61(3), 421–432.
[38] Lv X, Li S. Graphene oxide–crospolyvinylpyrrolidone hybrid microspheres for the efficient adsorption of 2,4,6-trichlorophenol, ACS Omega 2020, 5, 18862–18871.

Chapter 3
Derivatives from acetylene reacting with alcohol

The general formula of vinyl ether compounds is $R_1C(R_2)=C(R_3)OR_4$ (where R_1=H or alkyl; R_2=H, Me, Et or Cl; R_3=H, Me or Et; and R_4=alkyl or alkyl with functional groups). In these unsaturated ethers, vinyl double bond is directly connected to an ether bond. Due to the influence of an adjacent oxygen atom, the electron-rich double bond exhibits a high reactivity [1].

Vinyl ethers are commonly prepared by vinylation of the corresponding alcohol with acetylene in the presence of an alkali metal alkoxide (sodium alkoxide, potassium alkoxide, etc.) or an alkali metal hydroxide (NaOH and KOH) catalyst. The reaction is usually carried out at elevated temperature and pressure. The reaction equation is as follows:

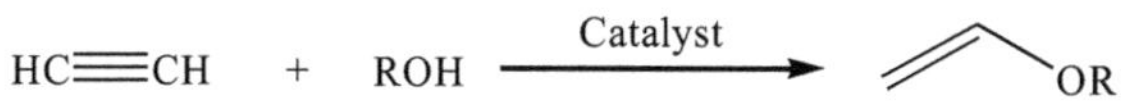

The reaction of acetylene with alcohol belongs to the nucleophilic addition process. First, the K^+ ion in the potassium salt complexes with the acetylene molecule to form a strongly basic alkoxy anion. The alkoxy anion will take the proton of the alcoholic hydroxyl group and form a new alkoxy anion. Then, under heat and pressure, the generated alkoxy anion attacks the alkyne carbon for nucleophilic addition; finally, the alkyne carbon anion takes protons from the alcohol to form an vinyl ether.

3.1 Methyl vinyl ether

Methyl vinyl ether (MVE), also known as methoxyethylene, vinyl methyl ether and 1-methoxyethylene, is the simplest chemical that contains both an alkene group and an alkyl group with an ether linkage. CAS Registry Number: 107-25-5. Molecular formula: C_3H_6O; molecular weight: 58.0791. The chemical structure is

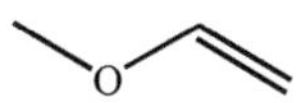

3.1.1 Properties of methyl vinyl ether

MVE is a colorless and flammable gas with a sweet and pleasant odor at room temperature. It is heavier than air and can diffuse to a relatively distant place at a lower place. MVE is soluble in ethanol, ether and acetone, but it is only slightly soluble in water. The physical properties of MVE are listed in Table 3.1. MVE is easily polymerized, so potassium hydroxide is often added as a polymerization inhibitor in the finished product.

https://doi.org/10.1515/9783110714999-003

Table 3.1: Physical properties of methyl vinyl ether.

Melting point/°C	−121.6	Flash point (open cup)/°C	−51°C
Boiling point/°C	5.5	Heat capacity ($C_{p,gas}$)/ J · mol^{-1} · K^{-1}	
Enthalpy of vaporization (20 °C)/ kJ · mol^{-1}	23.4	25 °C	79.50
Density (20 °C)/g · cm^{-3})	0.751	27 °C	79.91
Refractive index (20 °C)	1.3947	127 °C	99.16
Solubility in water (20 °C)/wt%	0.97		

3.1.2 Method for production of methyl vinyl ether

The industrial processes of MVE production include acetylene route (Reppe process) and acetal route according to the starting materials.

3.1.2.1 Acetylene route

Acetylene undergoes a nucleophilic addition reaction with methanol in the presence of a strong base to produce MVE:

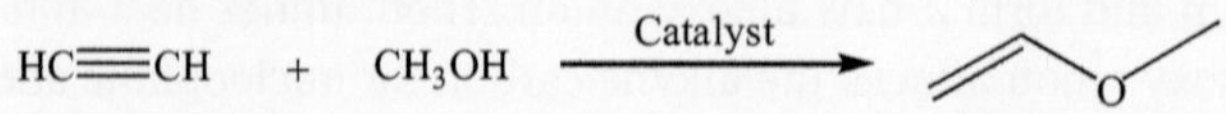

The original Reppe process was carried out at a temperature of 160–165 °C under a pressure of 2–2.2 MPa. The most commonly used catalyst for this reaction was an alkali metal hydroxide or an alkali metal alkoxide. Acetylene was usually diluted by nitrogen.

The Reppe process has the advantages such as low cost of raw material, one-step reaction and easy separation of the product after the reaction. However, the Reppe process requires high temperature and high pressure, leading to a safety problem of handling acetylene gas under such conditions. Moreover, the conversion of acetylene and yield of MVE are not high enough to be satisfactory. Therefore, the application of the process for industrial production of MVE is limited.

Improvements in this process are associated with the development of new catalyst, leading to mild reaction conditions. The most common way to develop a better catalyst is to increase the basicity of the catalyst which thereby enhances the activity. Dirk et al. [2] used a superbase system as the catalyst. In the system, potassium alkoxide and crown ether were dissolved in an ionic solvent such as an alkane. The catalyst system significantly improved the catalytic activity of potassium alkoxide, but the reaction still needed to be carried out under pressure. Recent advances in this field of superbase catalyst system was potassium hydroxide–dimethyl sulfoxide (KOH–DMSO), and the yield of MVE was more than 90%. The superbase catalyst system enabled the vinylation process of methanol or other lower alcohols operating under atmospheric pressure and at a temperature as low as 80–100 °C with no diluent of acetylene. The yield of MVE was almost quantitative [3, 4].

An inert solvent (glycol dimethyl ether) having a high solubility for acetylene was used for the vinylation of methanol. Acetylene gas was introduced into an autoclave containing KOH, methanol and glycol dimethyl ether, and the reaction was carried out at a temperature of 120 °C for 25 h. The conversion of methanol was 51.6% [5].

A British patent [6] described a process using a molten mixture of alkali metal hydroxides as the catalyst. An equimolar mixture of acetylene gas and vaporized methanol was passed through a molten mixture of 51 wt% of sodium hydroxide and 49 wt% of potassium hydroxide at a temperature of 222 °C. The conversion of methanol was 16% and the conversion of acetylene was 10%.

Heterogeneous catalysts were also developed [7]. Solid base catalysts such as KOH/C, NaOH/C, NaOH/CaO and KOH/MgO were used in a study of the reaction between acetylene and methanol to prepare MVE. The catalysts exhibited very high conversion initially but the activity was declining to less than 10% conversion after 24 h. The deactivation of the catalysts was caused by trace amounts of impurities such as water and carbon dioxide as well as the deposition of by-products on the catalyst surface.

An amorphous catalyst prepared from zinc oxide on silica was found to be an efficient catalyst with excellent catalytic life for many months [7]. An equimolar mixture of acetylene gas and vaporized methanol was introduced into a fixed-bed reactor filled with ZnO/SiO_2 catalyst. The reaction temperature was in the range from 200 to 270 °C. MVE is the main product, and significant amount of by-product (dimethoxyethane) was also generated by an equilibrium reaction of MVE and methanol to dimethoxyethane. A higher reaction temperature favored that increase of MVE content in the reaction mixture, while a lower reaction temperature favored the formation of dimethoxyethane.

3.1.2.2 Acetal route

MVE is prepared by pyrolysis of dimethylacetal at high temperature in the presence of solid catalyst:

$$CH_3CH(OCH_3)_2 \longrightarrow CH_3OCH{=}CH_2 + CH_3OH$$

In general, the pyrolysis of dimethylacetal is carried out in the vapor phase. Dimethylacetal was vaporized and introduced into a fixed-bed reactor filled with calcium phosphate catalyst. In the reactor, dimethylacetal split into MVE and methanol at 320 °C and under atmospheric pressure. The conversion of dimethylacetal was 95%, and the selectivity to MVE was more than 98%.

3.1.3 Uses of methyl vinyl ether

MVE is widely used in organic synthesis due to its active double bond properties. It can be used to synthesize glutaraldehyde, γ-pyran, γ-pyridine, β-lactam and other compounds [8].

MVE is a monomer for polymers including homopolymer and copolymers [9, 10]. The homopolymer of MVE, also known as PVM, poly(methyl vinyl ether) and poly(vinylmethyl ether), is a soft resin and is used as thickener, plasticizer, heat sensitizer and raw material for adhesives, which can be produced by bulk polymerization of MVE using boron trifluoride as an initiator. The polymerization process is as follows:

About 200 mL of boron trifluoride solution containing 3% boron trifluoride dihydrate in dioxane was added to 300 L MVE in a polymerizer equipped with a reflux condenser. After the strongly exothermic polymerization was initiated, 5,300 L of MVE (during 10 h) and 6 L boron trifluoride solution (during 16 h) were continuously introduced into the polymerizer. The heat of polymerization (1,400 kJ kg^{-1} monomer) was removed by jacket cooling and the reflux condenser. Three hours after the monomer addition was terminated, the temperature increased from 12 to 60 °C, and 2 h later to 90 °C. After the addition of boron trifluoride solution was finished, the cooling water was switched off. Two hours later, the stirrer was switched off. After a further 2 h, the polymerizer is degassed for 4 h. The polymer was converted into 50% solutions with solvents such as water, ethanol or toluene. The conversion of the monomer was approximately 97%. The polymer had a weight-average molecular weight (Mw) of 60,000 or a number-average molecular weight (Mn) of 20,000.

PVM is mainly manufactured by BASF under the trade name Lutonal M 40. Several product forms including soft resin, 50% aqueous solution and 70% ethanol solution are available. MVE can also be copolymerized with other monomers for the preparation of coatings, plasticizers, adhesives and a modifier for polystyrene resin. In particular, the copolymer of MVE and maleic anhydride (MA) is most widely used [11–15]. The copolymerization process is as follows.

MA was added to a solvent mixture of cyclohexane and ethyl acetate in a polymerizer equipped with an agitator, a reflux condenser and an inner temperature-controlling device. The polymerizer was purged of air and filled with nitrogen. The solution was heated to 70 °C and then an initiator solution (lauroyl peroxide dissolved in ethyl acetate) and MVE were added to the polymerizer in a dropwise manner. A slurry was obtained, which was filtered and dried under vacuum to obtain a white powdery copolymer.

The copolymer of MVE and MA is a straight chain and water-soluble polymeric electrolyte compound. Its physical and chemical properties are similar to those of PVP series. It is excellent in chemical stability, biocompatibility, tackiness, cohesiveness, water-holding property, peeling property and the like. Furthermore, they are nontoxic and harmless to the human body. Therefore, it is widely used in oral care, cosmetics and medicine and other fields. The copolymer is used as a dental material requiring excellent adhesion, durability and biocompatibility with the oral mucosa. It has a very strong ability to inhibit tartar, therefore it is applied in toothpaste and mouthwash formulations for preventing tartar, reducing dentine hypersensitivity, increasing the bioavailability of active substances and making the effect of oral care products more durable. In addition, the copolymer is used for preparation of a cross-linked hydrophilic coating on the surface of the substrate to improve the surface lubrication of the catheter. This modified interventional catheter can reduce friction and tissue damage, reduce protein adsorption, prevent thrombosis and bacterial adhesion, which has broad application prospects in the medical field.

Nanoparticles based on PVM/MA copolymer can encapsulate or incorporate a product of interest for use in the agricultural, cosmetic, food or pharmaceutical fields [16].

Currently, the largest manufacturer of MVE/MA copolymer in the world is Ashland (former ISP). It has developed a series of MVE polymer products under the trade name Gantrez. The production of the series of products is primarily by copolymerization of MVE and MA to form a series of Gantrez AN copolymers with different ranges of molecular weight:

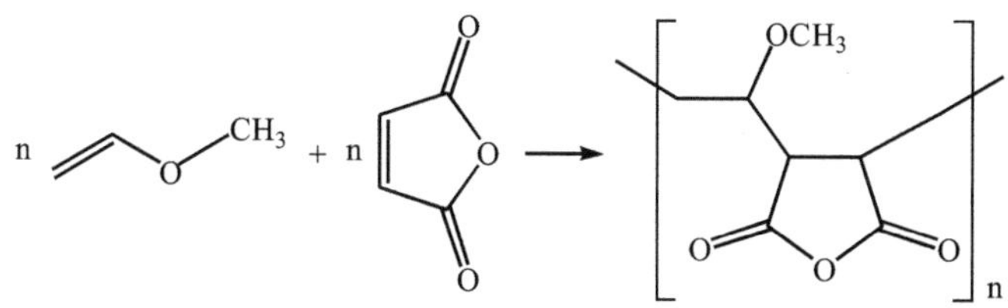

Then, Gantrez AN reacts with water, alcohol or calcium/sodium ions to produce Gantrez S, Gantrez ES and Gantrez MS, respectively. Figure 3.1 shows processes from Gantrez AN to Gantrez S, Gantrez ES and Gantrez MS.

GANTREZ AN copolymers are widely used as a thickener and flocculant, dental adhesive and excipient in oral tablets. In addition, they are also used for the controlled release of drugs. GANTREZ S copolymers are the free acid form of MVE/MA copolymer and used in toothpastes and mouthwashes as an excellent mucosal adhesive for the retention and prolonged delivery of antimicrobial agents and flavors. They also impart tartar control, anti-demineralization and dentin tubule flow reduction. GANTREZ MS is a mixed sodium and calcium salts of PVM/MA copolymers. It is used in denture adhesives, ostomy adhesives and in topical carriers for mucosal applications. GANTREZ ES copolymers are the half ester form of different alkyl chain lengths and molecular weights of MVE/MA copolymers. They are used in enteric film coating agents and in ostomy adhesives.

Other manufacturers of MVE/MA copolymers include BASF, Pantasote, Tenneco Chem., Du Pont and NKY Medical Group.

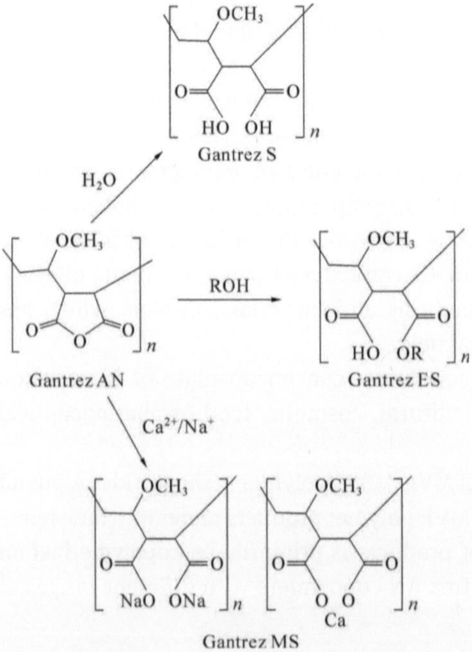

Figure 3.1: Processes from Gantrez AN to Gantrez S, Gantrez ES and Gantrez MS.

3.2 Ethyl vinyl ether

Ethyl vinyl ether, abbreviated as EVE, also known as vinamar, ethoxyethene, ethoxyethylene, vinyl ethyl ether, 1-ethoxyethene, ethyl ethenyl ether, 1-ethoxyethylene and ethyloxyethene, is a derivative of ethanol and acetylene. CAS Registry Number: 109-92-2. Molecular formula: C_4H_8O, molecular weight: 72.1057. The chemical structure is

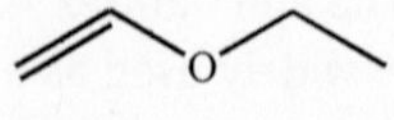

3.2.1 Properties of ethyl vinyl ether

EVE is a colorless and transparent liquid with a boiling point of 35.7 °C. It is miscible with various organic solvents such as acetone, benzene, diethyl ether, heptane, methanol and carbon tetrachloride. It is only slightly soluble in water. Under liquid–liquid equilibrium at 20 °C, the concentration of EVE in water is 0.9% by weight, while the amount of water dissolved in EVE is 0.2% by weight. The physical properties of EVE are summarized in Table 3.2. The vapor of EVE and air can form an explosive mixture. It is easy to burn and explode in case of open flame and high heat. The explosion limit (vol) is in the range from 1.7% to 28%. EVE should be stored in

Table 3.2: Physical properties of ethyl vinyl ether.

Melting point/°C	−115.3	Viscosity (20 °C)/mPa · s	0.2
Boiling point/°C	35.7	Flash point (open cup)/°C	−18
Enthalpy of vaporization (−35 °C)/ $kJ \cdot mol^{-1}$	29.5	Ignition point/°C	201.7
Density (20 °C)/ $g \cdot cm^{-3}$	0.7541	Vapor pressure (20 °C)/kPa	56.80
Refractive index (20 °C)	1.3754	Volume expansion coefficient (20 °C)	0.00165

a ventilated, low temperature and dry place, in particular, separately from oxidants and acids.

EVE is chemically active and reacts violently with halogens and strong oxidizing agents. Self-polymerization can occur due to the action of light, heat, peroxide and so on, and the polymerization reaction is rapidly increased with the elevation of temperature. It is stable to alkalis, but easily hydrolyzed to produce ethanol and acetaldehyde in the presence of inorganic acids. At low temperatures, EVE undergoes an addition reaction with hydrogen halide to give an α-halogenated hydrocarbon. Hydrogenation of EVE in the presence of Raney nickel catalyst generates diethyl ether.

3.2.2 Process for manufacture of ethyl vinyl ether

EVE can be prepared by reacting acetylene with absolute ethanol in the presence of an alkali catalyst:

$$HC{\equiv}CH + CH_3CH_2OH \xrightarrow{\text{Catalyst}} H_2C{=}\underset{H}{C}{-}O{-}CH_2CH_3$$

The most commonly used catalyst for the vinylation is an alkali metal hydroxide or an alkali metal alkoxide.

In China, three processes for the manufacture of EVE by acetylene route were used. The first one is a continuous process with homogeneous catalyst under high pressure. The advantages of this process are fast reaction rate and high conversion, but the disadvantages include requirement of high standard equipment, large energy consumption and easy to cause safety issues. The second one is a process with solid catalyst (heterogeneous catalyst) under atmospheric pressure, which is relatively simple in product separation and refining comparing with the process with homogeneous catalyst under high pressure, but has the disadvantages of low output, short service life of catalyst (about 110 h) and high requirement on the specification of carrier lime. The third one is a process with homogeneous catalyst under atmospheric pressure, which has the advantages of high output and good safety, but has the disadvantage of low conversion.

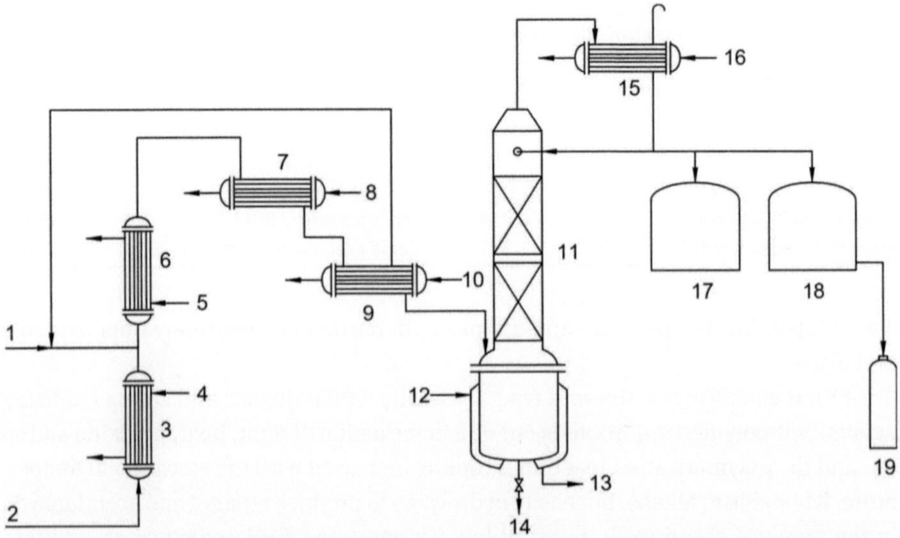

1.Acetylene; 2. Ethanol; 3. Evaporator; 4, 12. Steam; 5. Heat transfer oil; 6. Reactor; 7. The first condenser; 8. Cooling water; 9. The second condenser; 10, 16. Refrigerated brine; 11. Distillation tower; 13. Condensate drain; 14. Residue; 15. Condenser; 17. Storage tank; 18. Product tank; 19. Steel cylinder.

Figure 3.2: Process flow diagram for manufacturing ethyl vinyl ether by gas–solid catalysis under atmospheric pressure.

In the process with solid catalyst under atmospheric pressure, acetylene and ethanol vapor were introduced into a fixed-bed reactor and the vinylation reaction was carried out at a temperature as high as about 180 °C. EVE was produced with a concentration of about 70% in the outlet gas stream from the reactor. The fixed-bed reactor was charged with 4–5 mesh catalyst particles of potassium hydroxide supported on lime. The production process flow is shown in Figure 3.2.

The process with homogeneous catalyst under atmospheric pressure includes three steps. First, a hot catalyst solution was prepared by dissolving potassium hydroxide in absolute ethanol under stirring and heating. The weight of potassium hydroxide applied was 1.1–1.3 times that of ethanol, ensuring the temperature of the KOH solution to reach 140 °C. The hot catalyst solution was added to a bubble column reactor which had been preheated to 140 °C by the heat transfer oil circulating in the jacket. The height/diameter ratio of the reactor was 70:1. Acetylene and absolute ethanol solution containing 2–3% of potassium hydroxide were simultaneously and continuously introduced from the bottom of the reactor. A constant liquid level in the reactor was maintained by controlling the flow rate of the absolute ethanol solution containing 2–3% of potassium hydroxide. The reactor was operated at 140 °C and under atmospheric pressure. A conversion reached 88–90% based on acetylene. The EVE generated in the reaction and the unreacted ethanol were distilled from the top of the reactor, condensed by two condensers connected in series and collected in a crude product tank. The

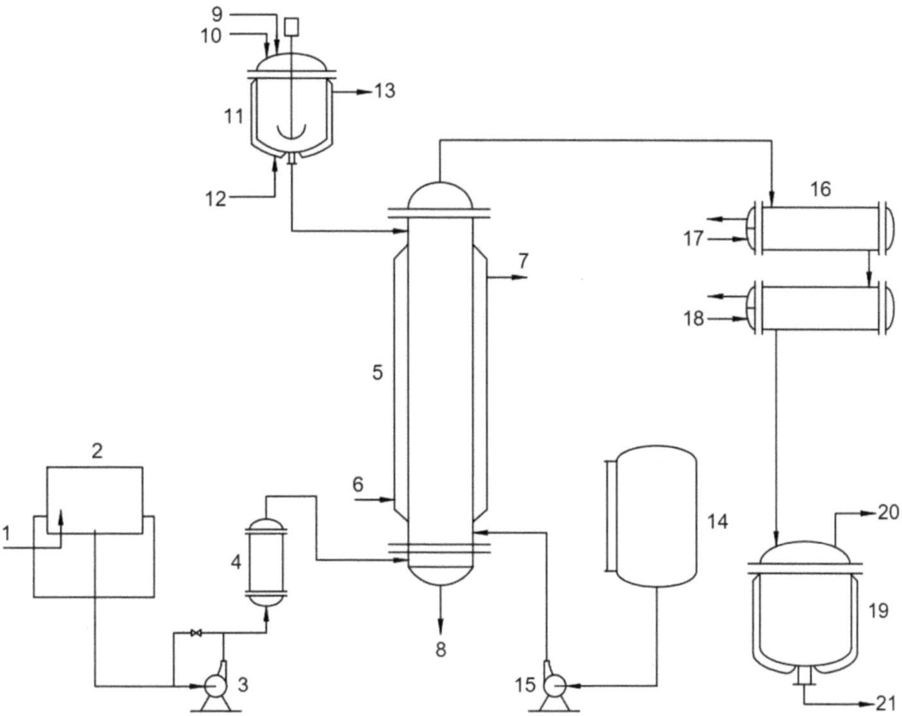

1.Acetylene; 2.Acetylene storage tank; 3. Acetylene pump; 4. Dryer; 5. Reactor; 6, 12. Heat transfer oil inlet; 7, 13. Heat transfer oil outlet; 8.drain; 9.Potassium hydroxide; 10. Ethanol; 11. Stirred kettle; 14.storage tank for ethanol solution; 15. ethanol solution pump; 16. Condenser; 17. Cooling water; 18. Refrigerated brine; 19. Crude product tank; 20. Acetylene recovery; 21 Crude product of ethyl vinyl ether.

Figure 3.3: Process flow diagram for manufacturing ethyl vinyl ether by homogeneous catalysis under atmospheric pressure.

unreacted acetylene was recycled. The concentration of EVE in the crude product was about 70–80%. The catalyst activity gradually decreased with the reaction time, therefore, it was necessary to regularly release part of the catalyst solution with declining activity from the reactor. The production process flow diagram is shown in Figure 3.3.

EVE mainly contains impurities such as ethanol, acetaldehyde, acetal and triethanolamine added to prevent hydrolysis, which need to be refined before use. The purification is carried out by extensive washing with water or dilute potassium hydroxide solution.

Peroxide may be generated in EVE if the latter has been stored for a long time. In this case, potassium iodide dissolved in dioxane can be applied to check if the ether contains peroxide. The peroxide in the ether can be washed by reducing agent solution of ferrous sulfate, sodium sulfite, stannous chloride and the like, dried by potassium hydroxide and calcium chloride, and finally refined by distillation in the presence of potassium hydroxide or metallic sodium.

3.2.3 Uses of ethyl vinyl ether

EVE is a monomer for polymers including homopolymer and copolymers. The homopolymer of EVE is a raw material for adhesives and paints. The copolymer of EVE and MA is widely used for pharmaceutical purposes as a thickening and suspending agent, denture adhesive and adjuvant for transdermal patches.

EVE is also used as an intermediate for fine chemicals such as sulfadiazine and the disinfectant glutaraldehyde. Vinyl ether can paralyze the central nervous system, and its anesthetic effect is twice as strong as ether, so it can be used as anesthetic and analgesic in medical treatment.

In addition, EVE is used in organic synthesis such as protection of hydroxyl group, transvinylation and cycloaddition reactions. EVE reacts with hydroxyl to form α-ethoxyethyl ether, which is one of the most commonly and convenient way to protect hydroxyl group. Usually, the reaction requires a strong acid catalyst such as trifluoroacetic acid, *p*-toluenesulfonic acid, and pyridinium *p*-toluenesulfonate. α-Ethoxyethyl ether is easily to complete the deprotection reaction under acidic conditions. In the presence of catalysts (such as mercury acetate, phosphoric acid and *p*-toluenesulfonic acid), EVE and allylic alcohols undergo transvinylation reaction to generate allyl vinyl ethers. This reaction is more significant because the products can undergo Claisen rearrangement to give unsaturated aldehydes. Moreover, EVE can also undergo a corresponding cyclization reaction in the presence of a Lewis acid catalyst.

3.3 *n*-Butyl vinyl ether

n-Butyl vinyl ether, abbreviated as BVE, also known as butyl vinyl ether, butoxyethene, butyl vinyl ether, vinyl butyl ether, vinyl *n*-butyl ether, butoxyethylene, ethenyl *n*-butyl ether and 1-butoxyethylene, is a derivative of *n*-butanol and acetylene. CAS Registry Number: 111-34-2. Molecular formula: $C_6H_{12}O$, molecular weight: 100.1589. The chemical structure is

3.3.1 Properties of *n*-butyl vinyl ether

BVE is a colorless, transparent and extremely flammable liquid. It is lachrymatory and irritates the respiratory system. BVE is miscible with various organic solvents such as benzene, diethyl ether, hexane and carbon tetrachloride. The solubility of BVE in water is 0.10% by weight, and the solubility of water in BVE is 0.09% by weight at 20 °C. The physical properties of BVE are summarized in Table 3.3.

Table 3.3: Physical properties of *n*-butyl vinyl ether.

Melting point/°C	−112.7	Heat capacity (25 °C)/ J · mol^{-1} · K^{-1}	231.8
Boiling point/°C	94.0	Vapor pressure (20 °C)/kPa	5.60
Enthalpy of vaporization/kJ · mol^{-1}		Density (20 °C)/g · cm^{-3}	0.7803
284 K	36.1	Refractive index (20 °C)	1.4017
320 K	35.2	Viscosity (20 °C)/mP · s	0.47
360 K	32.5	Critical temperature/K	540.5
367.1 K	31.58	Critical pressure/MPa	3.20
400 K	29.6	Volume expansion coefficient (55 °C)	0.00133

The carbon–carbon double bond of BVE is chemically active due to the influence of an alkoxy group and can undergo an addition reaction with a plurality of types of substances to form various derivatives. BVE does not undergo homogeneous polymerization alone, but it can be polymerized by a catalyst such as aluminum trichloride or boron trifluoride. It is stable to alkalis but readily hydrolyzes to form *n*-butanol and acetaldehyde under acidic conditions.

3.3.2 Process for manufacture of *n*-butyl vinyl ether

BVE is produced by vinylation of *n*-butanol with acetylene in the presence of a catalyst:

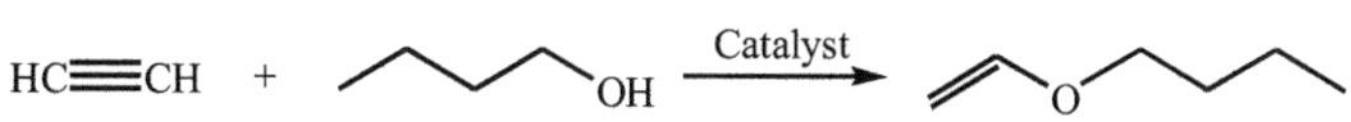

First, potassium hydroxide reacts with *n*-butanol to form alkoxide:

$$C_4H_9OH + KOH \rightleftharpoons C_4H_9OK + H_2O$$

Then, the alkoxide reacts with acetylene to obtain the potassium-substituted BVE:

$$HC \equiv CH + C_4H_9OK \longrightarrow KHC{=}CHOC_4H_9$$

The potassium-substituted BVE reacts with *n*-butanol to generate BVE and alkoxide:

$$KHC{=}CHOC_4H_9 + C_4H_9OH \longrightarrow H_2C{=}CHOC_4H_9 + C_4H_9OK$$

The alkoxide continues to react with acetylene.

The vinylation is generally carried out via gas–liquid reaction in a bubble column reactor, which has the advantages of simple structure, low cost and no sealing problems because of no mechanical stirring. In the reactor, acetylene gas passes through the liquid layer through the distributor in a bubbling manner, forming a phase interface necessary for the gas–liquid reaction. Moreover, the rising gas flow promotes the liquid phase to be in a turbulent state, thereby enhancing the chemical reaction, mass transfer and heat transfer in the bed.

The mass and heat transfer effect of the bubble column reactor is related to a number of factors, such as flow rate of acetylene gas, liquid phase viscosity and tower structure. When the apparent gas velocity is less than 0.05 $m \cdot s^{-1}$, the structure of the gas distributor determines the dispersion of the gas and the size of the bubbles, which in turn determines the gas holdup and the mass transfer coefficient of the liquid phase. When the apparent gas velocity is greater than 0.1 $m \cdot s^{-1}$, the bubble is mainly formed by the impact and friction between the gas stream and the liquid. The size and distribution of the bubbles mainly depend on the apparent gas velocity, but has little relationship with the structure of the gas distributor.

The flow state of the liquid phase in the reactor is related to the flow rate of acetylene gas. For the acetylene/*n*-butanol system at a temperature of 170 °C and under a pressure of 5 MPa, when the gas velocity is lower than 0.06 $m \cdot s^{-1}$, bubbles are uniformly distributed in the tower, and the bubble sizes are uniform, which float and rise regularly in the liquid phase. In this case, the fluctuation of the liquid surface in the bed is not large, and the axial back mixing is not serious, which is commonly called the operation state of stable zone. When the gas velocity is greater than 0.08 $m \cdot s^{-1}$, the bubbles merge and increase size rapidly in the column, and move irregularly in the liquid phase, stirring the liquid very intensely. At this time, the liquid layer is in a turbulent zone operation state with serious axial back mixing, and the residence time of the target product is prolonged, resulting in significant increase of side reactions and decrease of yield of the synthesis reaction.

Since the synthesis of BVE is a rapid reaction, it is very important to determine the apparent gas velocity of acetylene according to the actual situation. The gas–liquid reaction should be operated in the state of stable zone as much as possible to reduce axial back mixing and side reactions. In addition, if the residence time of the acetylene gas passing through the liquid layer in the tower is too short, the reaction is insufficient and the conversion is low; if the residence time is too long, the side reaction is increased and the yield of BVE is reduced. Therefore, the height of the liquid layer in the tower also affects the yield of BVE. It is suggested that the preferable residence time of acetylene in the liquid layer is in the range from 90 to 120 s [17].

There are also some shortcomings in this process. First, because the amount of exhaust gas generated by the bubble column is not equal in each time period, the adjustment of acetylene concentration is difficult, and it is difficult to meet the gas concentration index requirement. Second, the excessively high temperature of the synthesis reaction will increase the coking in the column, causing difficulty in slagging, affecting the effective cycle of the reaction and causing a decrease in the reaction yield.

A superbase catalyst system CsF–NaOH–DMSO was developed for preparation of BVE from acetylene and *n*-butanol at a temperature of 100 °C and under atmospheric pressure. After reaction for 3.5 h, the conversion of *n*-butanol was 99% and selectivity to isobutyl vinyl ether (IBVE) was 79% [18].

3.3.3 Uses of butyl vinyl ether

BVE is a monomer for homopolymerization and copolymerization to produce polyvinyl ethers, which are widely used in coatings, adhesives, additives, plasticizers and viscosity index improvers.

3.3.3.1 Homopolymer of *n*-butyl vinyl ether

Poly(*n*-butyl vinyl ether) (PB) homopolymer is an oil-soluble chain polymer, which is mainly used as modifier of synthetic resin materials to improve the aging resistance, gloss retention, wear resistance, antifouling and so on. It is also a good plasticizer, ink additive, and viscosity index improver.

Polymerization of BVE is carried out in the presence of an initiator such as stannous chloride, stannic chloride, aluminum chloride and ferric chloride:

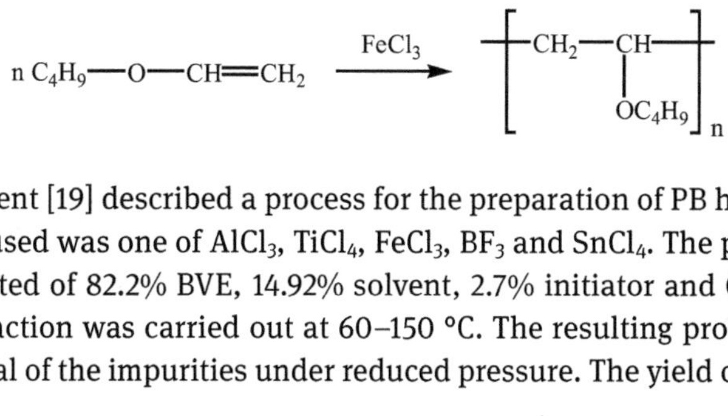

A Chinese patent [19] described a process for the preparation of PB homopolymers. The initiator used was one of $AlCl_3$, $TiCl_4$, $FeCl_3$, BF_3 and $SnCl_4$. The polymerization system consisted of 82.2% BVE, 14.92% solvent, 2.7% initiator and 0.18% capping agent. The reaction was carried out at 60–150 °C. The resulting product was purified by removal of the impurities under reduced pressure. The yield of PB was more than 93%.

PB is used as the first viscosity index improver in China. The trademark is T-601. The viscosity index improver for lubricating oil products is mainly used for blending multigrade oil, improving the viscosity–temperature performance, low-temperature startability and pumpability, reducing the oil consumption and wear of engines, and is mainly used in internal combustion engine oil, hydraulic oil, automatic transmission fluid and gear oil.

PB is a light yellow viscous liquid with a relative molecular weight of 9,000–12,000. The product can increase the viscosity of lubricating oil and improve the viscosity–temperature performance to meet the general requirements of oil products in winter and summer. T-601 has good shear stability and low temperature performance, but its thickening ability is not strong and its thermal stability is not high. The main types of viscosity index improvers are PB, polymethacrylate (PMA), polyisobutylene (PIB), ethylene–propylene copolymer (OCP), hydrogenated styrene–diene copolymer (HSD), polyacrylate and the like.

The viscosity-increasing performance of the viscosity index improver mainly depends on the molecular weight of the product. The viscosity-increasing effect increases with the increase of the molecular weight, and the length of the polymer chain (the number of carbons in the main chain) plays a decisive role, while the side chain has little effect on the viscosity-increasing performance. The order of viscosity enhancement of the viscosity index improver is as follows: HSD ≈ OCP > PB > PIB > PMA.

PB has a wide range of applications in industrial gear oils. Modern industrial gear oil is a kind of lubricating oil with high performance, low viscosity, versatility and modern additive formulation technology. The movement and power transmission of gear mechanism mainly depends on the mutual occlusion and relative movement of each pair of meshing tooth surfaces, which will inevitably cause friction. To avoid direct friction between the working surfaces of the gears, industrial gear oils (i.e., lubricants) are required to separate the working surfaces in order to maintain the efficiency of the gear mechanism and extend its service life. Industrial gear oil requires higher viscosity index, generally not less than 90 in medium load and not less than 95 in heavy load. However, the viscosity index of general intermediate base mineral oil cannot meet such high requirements, and the viscosity of most mineral oil has difficulty meeting the requirements of high viscosity oil products. Therefore, the viscosity index improver is often added to improve the viscosity index of gear oils to meet industrial requirements. The excellent shear stability of PB meets the requirements of industrial gear oil. If the viscosity of base oil is much different from that of finished oil, some low molecular polyisobutylene can be added to improve the viscosity, and then PB is added to improve the viscosity index, which can decrease the dosage of PB and reduce the cost of oil blending.

3.3.3.2 Copolymer of *n*-butyl vinyl ether and other monomers

BVE can be copolymerized well with other monomers, leading to some polymer materials with excellent properties [20]. For example, BVE is copolymerized with chlorotrifluoroethylene to produce a fluorocarbon coating with excellent properties. Room-temperature-curing fluorocarbon coating is a branch of fluorocarbon coatings which is widely used and rapidly developed. It is usually prepared by solution polymerization of fluorine-containing monomers (chlorotrifluoroethylene or vinylidene fluoride), vinyl ether (ester), hydroxyl vinyl ether and other functional monomers. The regular alternating structure between monomers will make the final paint film have excellent comprehensive properties.

3.4 Isobutyl vinyl ether

Isobutyl vinyl ether (IBVE), also known as vinyl isobutyl ether, 2-methyl-1-vinyloxypropane, 1-(vinyloxy)-2-methyl-propane, 1-(ethenyloxy)-2-methylpropane and isobutoxyethene, is a derivative of isobutanol and acetylene. CAS Registry Number: 109-53-5. Molecular formula: $C_6H_{12}O$, molecular weight: 100.1589. The chemical structure is

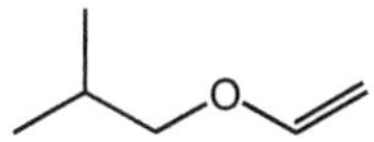

3.4.1 Properties of isobutyl vinyl ether

IBVE is a colorless, transparent and flammable liquid with an etheric aroma and low toxicity. Isobutyl vinyl ether is miscible with alcohols, ethers, esters and aromatic hydrocarbons. It is slightly soluble in water. The physical properties of IBVE are listed in Table 3.4.

Table 3.4: Physical properties of isobutyl vinyl ether.

Melting point/°C	−112.0	Heat capacity (25 °C)/ $J \cdot mol^{-1} \cdot K^{-1}$	231.8
Boiling point/°C	83.0	Vapor pressure (20 °C)/kPa	9.064
Enthalpy of vaporization/$kJ \cdot mol^{-1}$		Density (20 °C)/$g \cdot cm^{-3}$	0.7690
281 K	37.4	Refractive index (20 °C)	1.3960
355.4 K	30.7	Solubility in water (25 °C)/$g \cdot L^{-1}$	0.7

IBVE has a carbon–carbon double bond and is chemically active. It can react with a wide variety of substances to form various derivatives. IBVE is stable to alkalis but readily hydrolyzes to isobutanol and acetaldehyde under acidic conditions. Therefore, it is often stabilized by addition of a small amount of base such as triethanolamine.

3.4.2 Process for manufacture of isobutyl vinyl ether

There are two routes for production of IBVE: one uses acetylene as raw material, the other does not use acetylene as raw material [21–23]. In industry, acetylene route is mainly used for the production of IBVE.

IBVE is produced by vinylation of isobutanol with acetylene in the presence of a catalyst:

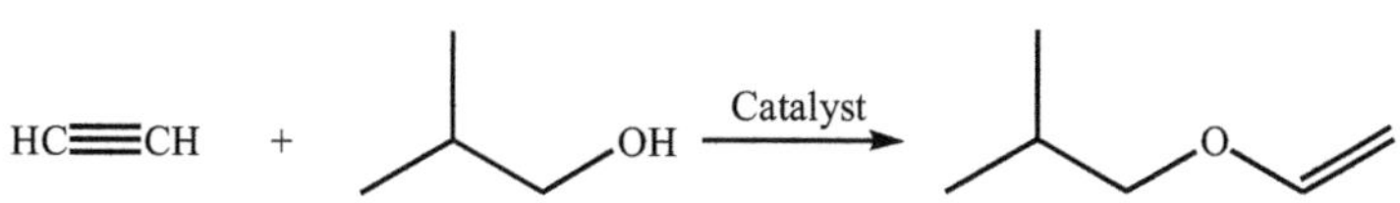

Commonly used catalysts for the acetylene process are alkali metal hydroxides (such as NaOH and KOH) and alkali metal alkoxides (such as sodium alkoxide and potassium alkoxide). The catalytic activity of KOH is greater than that of NaOH. The alkali metal elements of potassium and sodium can also be directly used as a catalyst, but the catalytic efficiency is low, and hydrogen gas is generated, which is disadvantageous to the reaction process, so that they are rarely used.

Potassium isobutoxide obtained by reacting potassium hydroxide with isobutanol has high catalytic activity and is usually used in production of IBVE in industry.

During the reaction, a high boiling polar solvent can be added to improve the reaction conditions. For example, *N,N*-diethylaniline was used as a solvent in the reaction system to increase the reaction temperature. Since the vinyl ether compound contains an unsaturated double bond, for the system where the product is easy to polymerize, a suitable polymerization inhibitor is often added to the system to prevent the side reaction of polymerization, thereby improving the yield of the product.

The processes for manufacture of IBVE by acetylene route include gas–liquid reaction process and liquid-phase reaction process. The states of acetylene in the reactors of the two processes are quite different.

3.4.2.1 Gas–liquid reaction process

Common reactors for production of IBVE by gas–liquid reaction process include autoclave, spray tower, bubble tower and loop reactor. In general, the autoclave is used for batch production, while the spray tower, bubble tower and loop reactor are used for continuous production.

Using dimethylaniline as a solvent and potassium isobutoxide as a catalyst, a stainless steel autoclave was applied for batch production of IBVE from acetylene and isobutanol [24]. The reaction temperature was increased and the vapor pressure of the alcohol was reduced in the solvent dimethylaniline with high boiling point. The reaction was carried out at a temperature of 160 °C and under a pressure of 0.5 MPa for 7 h. The conversion of isobutanol was more than 70%. This method is simple and basically has no tail gas emission. However, a large amount of toxic dimethylaniline needs to be separated and recovered after reaction, and the catalyst dosage is large (potassium isobutoxide is 10% of the mass of isobutanol), leading to a high production cost. In addition, the industrial scale-up of equipment is difficult.

IBVE is usually produced in a continuous process in industry, and the most common production units are spray tower and bubble tower.

In that spray tower, a liquid phase containing isobutanol and 8–16% of potassium hydroxide was added from the top of the tower, and gas mixture of acetylene and nitrogen was introduced into the tower from the bottom so as to make reverse contact of the gas and liquid phases for reaction at a temperature of 130–160 °C and under a pressure of 0.38–0.65 MPa. The IBVE generated in the reaction was discharged from the top of the tower along with the gas phase, condensed into liquid by a condenser, and then enters a receiving tank for phase separation. The unreacted acetylene and nitrogen were recycled. The conversion per pass is more than 57%, and the overall yield of IBVE was 95% [25]. The process is suitable for large-scale continuous production of IBVE, and has the advantages including stable product quality, high production efficiency and simple operation. However, there are some shortcomings in the reaction process, such as short gas–liquid contact time, large mass transfer resistance, incomplete reaction of acetylene gas and large energy consumption of gas circulation.

Compared to the spray tower, the conversion per pass of isobutanol in the bubble tower was improved to more than 80%, and the overall yield of IBVE was 98% [26]. The specific operation process is as follows:

Isobutanol solution containing 12–20% of potassium hydroxide was introduced into the bubble tower, and the liquid level in the tower was controlled to be 50–80%. After removal of the moisture by freeze-drying, acetylene gas was compressed and subjected to impurity removal by a drying tower, and then the acetylene and isobutanol protective gas are uniformly mixed through a nozzle from the bottom of the tower and then were introduced into bubble tower. The reaction was carried out at 125–150 °C and 0.3–0.6 MPa. The reaction solution at the tower bottom was sent to a heat exchanger by circulating pump for removal of the heat released by the reaction, and returned to the tower from the top for circulation. The formed small acetylene bubbles fully contacted with isobutanol in the bubble tower for reaction. The generated IBVE, unreacted acetylene gas and isobutyl alcohol were discharged from the top of the bubble tower, cooled and then collected in a receiving tank. The unreacted acetylene gas was recycled. During the reaction process, fresh raw material isobutanol and catalyst were continuously replenished by the pump, and the deactivated catalyst was regularly discharged.

The production of IBVE by bubble tower has the advantages including good safety and reliability, high conversion per pass, easy production control, simple operation and environmental protection. The disadvantage is that the amount of catalyst used is very large, increasing the production cost of IBVE.

A superbase catalyst system CsF–NaOH–DMSO was developed for preparation of IBVE from acetylene and isobutanol at a temperature of 100 °C and under atmospheric pressure. After reaction for five hours, the conversion of isobutanol was 98% and selectivity to IBVE was 76% [18].

3.4.2.2 Full liquid–phase cyclic reaction process

Li's research group in Xiamen University developed a process for continuous production of IBVE from acetylene and isobutanol by full liquid-phase cyclic reaction in a tubular reactor [27]. Pure acetylene gas was dissolved in the isobutanol solution containing 3% of potassium isobutoxide at low temperature and low pressure in advance. A solvent such as *N*-methyl pyrrolidone could be added to improve the solubility. Subsequently, the resulting solution saturated with acetylene was pressurized by a plunger metering pump and introduced into a tubular reactor for full liquid-phase reaction at a temperature of 155 °C and a pressure of 6–8 MPa. The residence time of the reaction solution in the reactor was 260 s. The outlet solution from the reactor was cooled and decompressed by a pressure relief valve, a small portion of the solution is discharged as the product, and the rest of the solution is circulated back to an absorber to absorb acetylene together with the feed liquid (isobutanol and catalyst), and then the solution is pressurized and introduced into the tubular reactor for continuous reaction. At steady state, the conversion of isobutanol reached 77.5%, and the selectivity of IBVE was up to 97.3%.

3.4.3 Uses of isobutyl vinyl ether

IBVE is a polymerization monomer for the production of poly (isobutyl vinyl ether) homopolymer [28]. It can also be copolymerized with other unsaturated compounds to prepare materials with special functions, such as polyvinyl chloride/IBVE resin, which is widely used in coatings, inks and surface protection materials.

3.4.3.1 Homopolymer of isobutyl vinyl ether

Poly(isobutyl vinyl ether) is used for plasticizer and as raw material for adhesives and paints.

IBVE polymerizes slowly in the presence of free-radical initiators such as azobisisobutyronitrile, peroxides or redox systems to form low-molecular-weight polymer of no commercial importance. The monomer contains an electron-rich double bond and is easy to form a stable carbocation. Therefore, it does not undergo an anionic polymerization reaction but is prone to cationic polymerization. With cationic initiators, polymerization of IBVE has a fast reaction rate and high conversion. However, the cationic polymerization reaction is easily inhibited by water, so it should be carried out in a dry environment as far as possible. Cationic initiators such as boron trifluoride, stannic chloride and aluminum chloride are usually used for the polymerization of IBVE. The polymerization process is similar to that of poly(methyl vinyl ether).

3.4.3.2 Preparation of vinyl chloride/isobutyl vinyl ether copolymer

Vinyl chloride/isobutyl vinyl ether copolymer (VC-IBVE resin) was first developed by BASF in Germany. The copolymer was obtained by emulsion copolymerization of 75% vinyl chloride and 25% IBVE. Chlorine content in the copolymer is about 44%. The appearance of the copolymer is white powder, and the particle size is generally in the range from 0.4 to 1.6 μm.

VC-IBVE resin has good resistance to water, stain, light, aging and chemical substances, and it is nontoxic. Due to the introduction of flexible isobutyl ether side chains in the resin structure, it has good internal plasticity, flexibility, miscibility and pigment wettability. Furthermore, as a chlorine-containing polymer, it also has excellent flame-retardant properties. Compared to traditional chlorinated rubber and chlorinated polyethylene, VC-IBVE resin has better adhesion to substrates such as metal materials. Currently, VC-IBVE resin has been widely used in anti-corrosion coatings, flame retardant coatings and ink binders [29–31].

The synthetic methods of VC-IBVE resin mainly include emulsion polymerization and micro-suspension polymerization in the presence of a radical initiator.

The emulsion polymerization technology for production of VC-IBVE resin was first developed and applied by BASF in Germany. In China, some companies such as Jiangsu Liside Chemical Co., Ltd. also use emulsion polymerization to produce VC-IBVE resin. In the emulsion polymerization, water phase is used as the continuous phase, and the polymerization is carried out at temperatures of 30–60 °C. The reactivity ratio of vinyl chloride is 2.19, which is much higher than that of IBVE (0.04). The homopolymerization of IBVE hardly occurs when the two monomers are copolymerized, and the self-polymerization tendency of vinyl chloride is much greater than the copolymerization tendency. In order to obtain the copolymer with uniform composition and good solubility in organic solvents, the vinyl chloride was added to the mixture of IBVE and vinyl chloride in a programmed manner during the polymerization process. Sodium lauryl sulfate was used as an emulsifier, and potassium persulfate as an initiator. The conversions of both monomers were more than 98%. The resulting copolymer was isolated by coagulation with a primary aliphatic alcohol or a water-soluble salt of a potassium cation.

If potassium persulfate alone is used as the initiator, the reaction takes a long time. The polymerization can be accelerated by using redox initiator system consisting of potassium persulfate (or ammonium persulfate) and sodium sulfite (or sodium bisulfite). The molecular weight of the copolymer increased as the polymerization temperature decreased.

In the micro-suspension polymerization process for production of VC-IBVE resin, the conversion is controlled at 65–80%. The unreacted IBVE and vinyl chloride are recovered and recycled. The main production formulation consists of vinyl chloride, IBVE, water, oil-soluble initiator, dispersant and the third monomer. The dosage of each component is shown in Table 3.5. The third monomer, such as acrylic acid, MA and vinyl neodecanoate, is applied for modification of the copolymer. In general, the micro-suspension polymerization is carried out at a temperature of 40–60 °C for 3–15 h.

Table 3.5: Micro-suspension polymerization formula for VC-IBVE resin.

Composition	Dosage/%	Remarks
Vinyl chloride	65–75	
Isobutyl vinyl ether	25–35	
The third monomer	0.03–5	Acrylic acid, maleic anhydride and vinyl neodecanoate
Water	100–500	

Table 3.5 (continued)

Composition	Dosage/%	Remarks
Initiator	0.5–1.5	Organic peroxide or azo initiators, such as dibenzoyl peroxide, azobisisobutyronitrile, *tert*-butyl peroxide and *tert*-butyl peroxide neodecanoate
Dispersant	0.05–5	Hydroxypropyl cellulose
Emulsifier	0.01–8	Anionic alkyl sulfate and alkyl sulfonate, nonionic polyoxyethylene ether and fatty acid ester

VC-IBVE resin is mainly produced by BASF. Its production started late in China, but has developed rapidly in recent years. Table 3.6 lists the main manufacturers of VC-IBVE resin, their capacities and technology used in China.

Table 3.6: Production status of VC-IBVE resin in China.

Manufacturer	Capacity/(t/a)	Technology
Jiangsu Jiangyin Huitong Fine Chemical Co., Ltd.	2,500	Micro-suspension polymerization
Hangzhou Electrochemical Group Co., Ltd.	3,000	Emulsion polymerization
Jiangsu Liside Chemical Co., Ltd.	4,000	Emulsion polymerization
Jiangsu Xuzhou Donghong Chemical Co., Ltd.	2,000	Emulsion polymerization and micro-suspension polymerization
Zhejiang Jinhua New Material Co., Ltd.	2,000	Micro-suspension polymerization

3.4.3.3 Reactive diluent

Because vinyl ether compounds have the advantages of low toxicity, low viscosity and easy synthesis, they can be used as safe and efficient reactive diluents for the production of epoxy resins, acrylic resins and unsaturated polyesters.

Reactive diluent is an important component of radiation-curing coatings, which is used to adjust the viscosity of coatings and the final properties of the film. Commonly used acrylate monomers can also be used as reactive diluents, but they are well-known carcinogens. Vinyl ether has low toxicity, no irritation to skin, low volatility and little pollution in construction. Therefore, vinyl ethers used as reactive diluents in UV-curing coatings have broad application prospects.

Zhou et al. [32] developed a kind of UV-curing screen-printing ink containing vinyl ether by copolymerizing vinyl ether with bisphenol A epoxy acrylate, phenolic epoxy acrylate and polyfunctional reactive monomer, suggesting that it was feasible to use vinyl ether as reactive monomer in epoxy acrylate UV-curing system.

IBVE, vinyl ethyl ether, chlorotrifluoroethylene and ethylene glycol monoallyl ether were used to prepare FEVE fluorocarbon resin elastic emulsion by emulsion polymerization in the presence of composite of anionic and nonionic emulsifiers. The coating film of the emulsion has good elongation at break and excellent water and alkali resistance, and can be used as a resin for a highly weather-resistant and environment-friendly elastic building exterior wall coating.

3.4.3.4 Copolymer of isobutyl vinyl ether and maleic anhydride

The IBVE-MA copolymer is also a material with good application prospects [33–35], and the copolymer is mainly used as a polymer additive in following two aspects.

First, the copolymer can be mixed with other materials for modification. It is reported that the polyvinyl alcohol is blended with IBVE-MA copolymer, and the modified material has better water resistance and mechanical properties. Polyvinyl alcohol is a kind of material with good properties, such as biodegradability, gas barrier property, transparency, antistatic property and organic solvent resistance. However, polyvinyl alcohol is a water-soluble polymer, which results in poor moisture resistance and easy moisture absorption. Moreover, polyvinyl alcohol has high rigidity and insufficient flexibility. Therefore, it needs to be modified as a packaging material. IBVE-MA copolymer could markedly improve flexibility of polyvinyl alcohol, and the elongation of the polyvinyl alcohol could improve to 300% after 2% of IBVE-MA copolymer was added to the pure polyvinyl alcohol.

Second, the copolymer of IBVE and MA has been used as emulsifiers in other production processes. Chlorpyrifos, an alternative to highly toxic pesticides such as methamidophos, has been widely used in agricultural and sanitary pest control. However, chlorpyrifos has a short duration of action on plant leaves due to its poor light stability. Moreover, the main dosage form of chlorpyrifos is an oily emulsion containing a large amount of organic solvent, which is easy to cause environmental pollution and waste of resources. One method to overcome these shortcomings is microencapsulation of chlorpyrifos. Pesticide microcapsules are the most promising types. They have the advantages of controlled slow release, long-lasting effect and high efficacy. IBVE-MA copolymer is used as an emulsifier to prepare pesticide sustained-release capsules [35]. The prepared microcapsules have regular and compact surface morphology and good performance, having good application prospects.

3.5 4-Hydroxybutyl vinyl ether

4-Hydroxybutyl vinyl ether, abbreviated as HBVE, also known as 1,4-butanediol vinyl ether, 4-vinyloxybutanol, butanediol monovinyl ether, tetramethylene glycol monovinyl ether and vinyl 4-hydroxybutyl ether, is a derivative of 1,4-butanediol and acetylene. CAS Registry Number: 17832-28-9. Molecular formula: $C_6H_{12}O_2$, molecular weight: 116.16. The chemical structure is

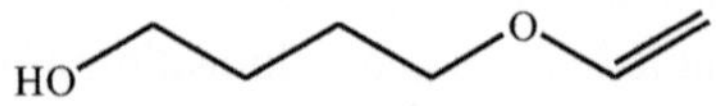

3.5.1 Properties of 4-hydroxybutyl vinyl ether

HBVE is a colorless or light yellow transparent liquid with a slight odor similar to ether. It is miscible with alcohols, ethers, esters and aromatic hydrocarbons. The physical properties of HBVE are listed in Table 3.7.

Table 3.7: Physical properties of hydroxybutyl vinyl ether.

Melting point/°C	−33.0	Density (20 °C)/g · cm^{-3}	0.943
Boiling point/°C	189	Flash point (open cup)/°C	88
Vapor pressure/kPa		Viscosity (20 °C)/mP · s	5.9
20 °C	0.03	Refractive index(20 °C)	1.4445
95 °C	2.67	Solubility in water (20 °C)/g · L^{-1}	75

3.5.2 Process for manufacture of 4-hydroxybutyl vinyl ether

HBVE is prepared by reacting acetylene and 1,4-butanediol in the presence of potassium hydroxide and potassium alkoxide catalyst:

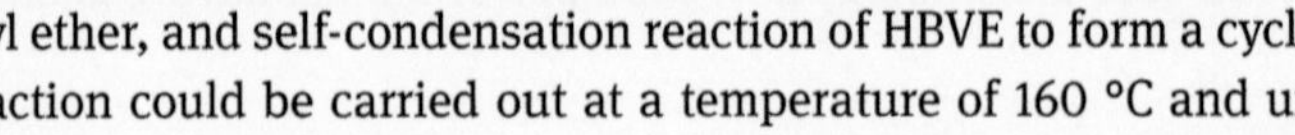

The conversion of raw materials is affected by factors such as reaction time, reaction temperature, reaction pressure and catalyst dosage. Since 1,4-butanediol contains two hydroxyl groups and the product contains active vinyl ether functional groups, some side reactions will occur. For example, further reaction of HBVE with acetylene to generate divinyl ether, and self-condensation reaction of HBVE to form a cyclic acetal.

The reaction could be carried out at a temperature of 160 °C and under atmospheric pressure, but the reaction was slow and conversion of 1,4-butanediol was low. Using a high concentration of potassium alkoxide catalyst up to 10–15% increased the conversion to 33–40%. However, the presence of large amounts of catalyst increased the production cost and also caused an increase in side reactions.

A superbase catalytic system CsF–NaOH was developed for the preparation of HBVE at a temperature of 138–140 °C and under initial acetylene pressure of 1.0–1.2 MPa. The conversion of 1,4-butanediol was 100% and the total yield of the vinyl ethers was 80% at catalyst dosage (7 mol% CsF and 7 mol% NaOH based on 1,4-butanediol) and after reaction for 3 h [36].

Some researchers also developed heterogeneous catalysts for the reaction of acetylene and 1,4-butanediol to obtain HBVE. Potassium hydroxide was supported on carriers such as aluminum oxide, molecular sieves, silica gel and zirconia. The conversion of 1,4-butanediol over these catalysts were much lower than that by homogeneous catalysis, and the supported catalysts usually deactivated after several cycles.

3.5.3 Uses of 4-hydroxybutyl vinyl ether

The structure of the HBVE is a vinyl double bond directly connected with an ether bond, and due to the influence of an adjacent oxygen atom, the double bond is an electron-rich double bond and shows higher reactivity. HBVE also has a hydroxyl, and can react with a series of resins. It has wide application in fluororesin and coatings.

3.5.3.1 HBVE as comonomer for fluororesin

Homopolymers and copolymers of fluorine-containing monomers can only be used for high-temperature thermoplastic coatings, and they have poor wettability to pigments and fillers as well as weak adhesion on the substrate surface. Therefore, some non-fluorine-containing monomers with polar cross-linking groups are usually added for copolymerization during the synthesis of fluororesin to ensure the compatibility of the resin and other additives, and these materials can be cross-linked and cured at room temperature. For example, fluorine-containing monomers, functional monomers and HBVE are subjected to solution polymerization in an autoclave, leading to fluororesin with excellent resistance to weather, oil, stain, water, solvent, chemical substances, tensile toughness, and very low friction coefficient, so that it has a wide range of uses in the fields of special anti-corrosion coatings, wear-resistant coatings and super-weatherable building coatings [37, 38].

At present, fluororesin materials are widely used in aerospace, industrial engineering, textile finishing, special protective coatings, microelectronics and other fields.

3.5.3.2 Application in coatings

With the development of science and technology, new coatings are also constantly developed. The scientific application of advanced technologies such as surface activation cross-linking technology, nanotechnology and supercritical fluid has better promoted the rapid development and application of new fluorocarbon coatings, and the enhancement of environmental awareness has led the development direction of new fluorocarbon coatings to green and environmental protection. For example, the cathodic electrophoretic coating prepared by copolymerization of fluoroolefin, vinyl ether, carboxyl-containing compound and water-soluble amino resin is a good waterborne fluoropolymer coating, which not only has the performance of pollution resistance and corrosion resistance of fluorine-containing materials, but also has good water solubility and construction performance, safety and environmental protection.

Reactive diluent is an important part of radiation-curing coatings, which is used to adjust the viscosity and the final properties of the coating. At present, the commonly used active diluents are acrylic esters, which have high volatility, low flash point, irritating odor and high toxicity, and many acrylic monomers are well-known carcinogens, so the environmental pollution problems caused by the use of acrylic monomers are difficult to overcome. Vinyl ethers, which have advantages including low toxicity, low viscosity and easy synthesis, is an excellent active diluent and used in paints, inks, adhesives, pressure-sensitive adhesives, surface modifiers and molding materials.

With the development of solvent-free coatings, the use of HBVE will continue to expand. Like free-radical UV-curing coatings, cationic UV-curing coatings are solvent-free coatings, which have the advantages including saving energy, reducing air pollution, fast curing speed, less land occupation, suitable for automatic assembly line coating and especially suitable for the coating of substrates that cannot be heated [39]. Modified epoxy resin UV-curing coating using HBVE as a reactive diluent has the advantages of fast curing, high gloss, good abrasion resistance, strong adhesion, excellent flexibility and low viscosity, which have been applied in the coating of beverage cans [40].

3.6 Diethylene glycol vinyl ethers

Diethylene glycol vinyl ether are also common vinyl ether compounds, including diethylene glycol monovinyl ether (DEGMVE) and diethylene glycol divinyl ether (DEGDVE).

DEGMVE is a derivative of diethylene glycol and acetylene. CAS Registry Number: 929-37-3. Molecular formula: $C_6H_{12}O_3$, molecular weight: 132.1577. The chemical structure is

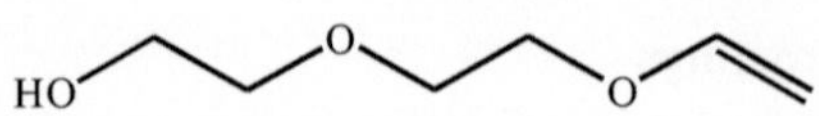

DEGDVE is also known as bis[2-(vinyloxy)ethyl] ether, divinylcarbitol and 1,1′-[oxybis(ethyleneoxy)]diethylene, CAS Registry Number: 764-99-8. Molecular formula: $C_8H_{14}O_3$, molecular weight: 158.1950. The chemical structure is

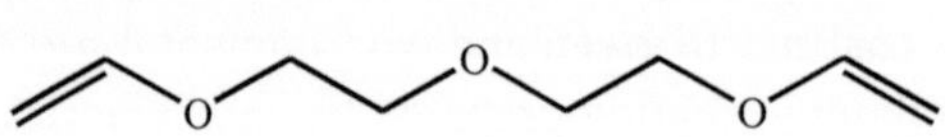

3.6.1 Properties of diethylene glycol vinyl ethers

DEGMVE is a colorless and transparent liquid with a slight odor similar to diethyl ether. It is soluble in water and some organic solvents such as alcohol and ether. The physical properties of DEGMVE are listed in Table 3.8.

Table 3.8: Physical properties of diethylene glycol monovinyl ether.

Melting point/°C	−50.4	Density (20 °C)/g · cm^{-3}	1.0267
Boiling point/°C	208	Flash point (open cup)/°C	82.8
Vapor pressure (80 °C)/kPa	0.3	Refractive index (20 °C)	1.4480
Vapor pressure (90 °C)/kPa	0.6		

DEGDVE is a colorless and transparent liquid with a boiling point of 198 °C under atmospheric pressure. It is miscible with alcohol, ether, ester and aromatic hydrocarbon, but only slightly soluble in water. The physical properties of DEGDVE are listed in Table 3.9.

Table 3.9: Physical properties of diethylene glycol divinyl ether.

Melting point/°C	−26.4	Density (20 °C)/g · cm^{-3}	0.9712
Boiling point/°C	198	Flash point (open cup)/°C	71.1
Vapor pressure (89 °C)/kPa	1.3	Refractive index (20 °C)	1.4460
Enthalpy of vaporization (78 °C)/kJ · mol^{-1}	50.0		

Both DEGMVE and DEGDVE are chemical active and easy to be polymerized, and a small amount of NaOH is often added as a polymerization inhibitor.

3.6.2 Production process of diethylene glycol vinyl ethers

The synthetic methods of diethylene glycol vinyl ethers mainly include acetylene route, acetal pyrolysis, dehydrohalogenation and transvinylation. The acetylene route is the only method for industrial production of diethylene glycol vinyl ethers.

3.6.2.1 Reaction

Acetylene reacts with diethylene glycol in the presence of a catalyst such as potassium hydroxide or potassium diethylene glycol to yield DEGMVE and DEGDVE:

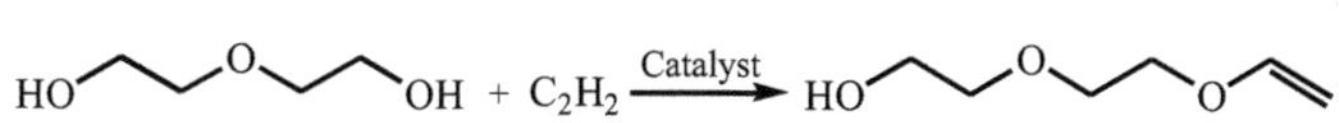

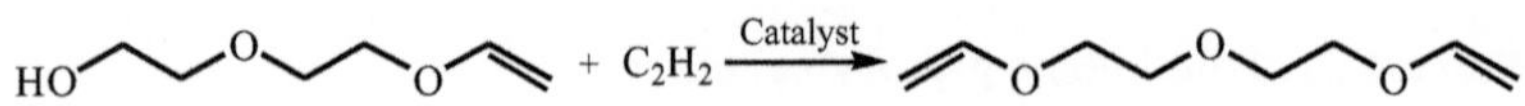

Acetylene route has the advantages including simple process, cheap and easily obtained raw materials, low production cost and less by-products. But the reaction needs to be carried out at elevated temperature and pressure, and acetylene gas is flammable and explosive.

Zhang et al. [41] prepared diethylene glycol vinyl ethers in a packed-bed reactor at a temperature of 175 °C and under a pressure of more than 0.6 MPa. The catalyst used was potassium diethylene glycol prepared from metal potassium, and the potassium concentration was 2% based on the mass of diethylene glycol. The gas–liquid reaction of acetylene and diethylene glycol was continuously carried out at acetylene space velocity of 0.3–0.4 h^{-1}. The conversion of the diethylene glycol reached 75%, and the overall yield of diethylene glycol vinyl ethers was 72%.

Zhao et al. [42] used a bubbling-bed reactor to carry out the gas–liquid reaction of acetylene and diethylene glycol for batch preparation of diethylene glycol vinyl ethers under atmospheric pressure. Potassium diethylene glycol was used as the catalyst; the catalyst concentration was 10%. After reaction for 8 h at a temperature of 175 °C and acetylene flow rate of 35 $mL \cdot min^{-1}$, the conversion of diethylene glycol was 63.7% and the total yield of diethylene glycol vinyl ethers was 61.2%.

Jiang et al. [43] compared the catalytic activity of potassium diethylene glycol and potassium hydroxide, and the results showed that the catalytic activity of potassium diethylene glycol was higher than that of potassium hydroxide. The conversion of diethylene glycol was 60% after reaction for 8 h at a temperature of 175 °C and a catalyst dosage of 8% based on the mass of diethylene glycol.

Li's research group in Xiamen University developed a process for continuous production of diethylene glycol vinyl ethers from acetylene and diethylene glycol by full liquid-phase cyclic reaction in a tubular reactor [44]. Pure acetylene gas was dissolved in the diethylene glycol solution containing 4% of potassium diethylene glycol at low temperature and low pressure in advance, and then the resulting glycol solution saturated with acetylene was pressurized by a plunger metering pump and introduced into a tubular reactor for full liquid-phase reaction at a temperature of 175 °C and a pressure of 6 MPa. The residence time of the reaction solution in the reactor was 175 s. The outlet solution from the reactor was cooled and decompressed by a pressure relief valve, a small portion of the solution was discharged as the product, and the rest of the solution was circulated back to an absorber to absorb acetylene together with the feed liquid (diethylene glycol and catalyst), and then the solution was pressurized and introduced into the tubular reactor for continuous reaction. At steady state, the conversion of diethylene glycol reached 76.03%, and the total yield of diethylene glycol vinyl ethers was up to 74.13%, in which the yield of DEGMVE was 59.03%, and the yield of DEGDVE was 15.10%. The reaction of diethylene glycol and acetylene was in good agreement with the first-order reaction kinetics. Because of no existence of gas-phase acetylene in the tubular reactor, the danger of flammable and explosive acetylene in the gas phase under high temperature and high pressure has been overcome.

3.6.2.2 Separation

The reaction solution discharged from the reactor was a mixture of diethylene glycol, potassium diethylene glycol, DEGMVE and DEGDVE. Therefore, a separation process was required so as to obtain pure diethyleneglycol monovinyl ether and pure diethyleneglycol divinyl ether, respectively.

Since the boiling points between DEGMVE and DEGDVE are not much different, fractional distillation requires a high column with larger number of theoretical plates. However, DEGMVE and DEGDVE forms an azeotrope, and pure DEGDVE cannot be obtained from the column top regardless of the reflux ratio and the number of theoretical plates.

The reaction mixture was subjected to distillation under vacuum. The azeotrope of DEGMVE and DEGDVE was distilled off at 133 °C and 6 kPa. The mass fraction of DEGDVE in the azeotrope was about 25%. Subsequently, the temperature was increased to 141 °C to distill off the residual DEGMVE.

The separation of azeotrope of DEGMVE and DEGDVE was performed by addition of potassium hydroxide in the azeotrope for binding DEGMVE, and then the mixture was subject to distillation for pure DEGDVE [45].

Another separation method for the reaction mixture is extractive distillation [46].

3.6.3 Uses of diethylene glycol vinyl ether

DEGMVE has good water solubility because of a hydroxyl group in the molecular structure. It can change the properties of the polymer by copolymerization with another monomer such as an epoxy compound. DEGDVE has two vinyl groups and is often used as a cross-linking agent in a polymerization system, making the cross-linked polymer to be an insoluble matter with spatial three-dimensional structure. In addition, both DEGMVE and DEGDVE can be used as reactive diluents in UV-curing cationic coating. Reactive diluent is an important component of radiation-curing coatings. Compared to traditional acrylate, diethylene glycol vinyl ethers have low toxicity, high flash point, no odor and no skin irritation, and they are safer and more efficient. Because of its low viscosity and good dilution performance, diethylene glycol vinyl ethers can make the coating completely solvent-free, so they are often used as active diluents for UV-curing coatings [47, 48].

3.6.3.1 Application of diethylene glycol monovinyl ether

DEGMVE has a hydrophilic property due to the presence of its side chain hydroxyl group, and can be homopolymerized into a hydroxyl group-containing polymer by cationic polymerization or radical polymerization. The polymer can be dissolved or swelled in water to form a solution or a dispersion, and has good properties such as stability and water retention, and has wide applications in fields such as biomedicine and surface protection materials. In addition, DEGMVE can be used as a comonomer for modification the polymer properties such as high tensile strength, high anti-contamination properties or environmentally sensitive properties.

The homopolymer of DEGMVE can be prepared by cationic polymerization. Since C=C of diethylene glycol vinyl ether will form p–π conjugate with oxygen having orphan electrons in the ortho position, the C=C forms an electron-rich structure, leading to a stable carbocation and cationic polymerization. In fact, the cationic polymerization of DEGMVE tends to form a liquid oligomer, mainly because the cationic polymerization occurs while the hydroxyl group on the DEGMVE reacts with the carbon–carbon double bond of the vinyl to form an acetal. Nowak [49] makes the hydroxyl group unaffected during the polymerization by continuously addition of a peroxide initiator. In addition, the homopolymer can also be synthesized by a radiation polymerization technique, that is, a polymerization reaction occurs when a gamma ray of ^{60}Co is used as a radiation source to irradiate the monomer to generate ions or radicals to form an active center, and the hydroxyl group remains in the reaction process [50].

DEGMVE can be used for preparation of an anti-protein adsorption material through plasma polymerization, and the resulting nanoscale functional coating has the characteristics of high controllability, good stability and excellent physical properties. DEGMVE is placed in a plasma reactor, and a plasma is formed by gas molecules under the action of an electric field to cause polymerization, thereby attaching an anti-protein adsorption layer to the substrate [51]. Proteins are amphiphilic macromolecules, which are easily adsorbed and contaminate materials. The hydration of DEGMVE forms a tight water film around the poly(diethylene glycol monovinyl ether) chain, and results in a large hydration repulsive force, preventing the adsorption of proteins on the surface, which is known as "water barrier effect" [52]. Due to the important application value of anti-protein adsorption materials in biomedical materials, surface protective coatings and other fields, they have gradually become the research hotspot.

DEGMVE is copolymerized with *N*-vinylpyrrolidone to form a chemically stable rubber-like polymer by emulsion polymerization, which has very strong tensile strength under the action of a vulcanizing agent [53]. In addition, DEGMVE reacts with alkyl bromide to synthesize a new monomer for preparation of a comb-like polymer by cationic polymerization. Due to the specificity of the side chain, the material has special phase transition behavior, and the heat storage value of the material reaches 180 J g^{-1}, which can be used as an excellent phase change material [54].

DEGMVE can also be used as a deactivator in the reaction of olefin metathesis [55]. The olefin metathesis reaction has been widely used in recent years. The new ruthenium carbene catalyst enhances the stability and reactivity of the reaction, but also causes some side reactions. The use of DEGMVE as a deactivator for the olefin metathesis reaction allows the ruthenium carbene catalyst to be rapidly deactivated at room temperature within 10 min.

3.6.3.2 Application of diethylene glycol divinyl ether

DEGDVE has two vinyl groups with adjacent oxygen atoms, and can be well compatible with other monomers in the absence of an emulsifier. It has a plurality of ether bonds, so that the polymerization product has small rotational steric hindrance and good flexibility. DEGDVE has attracted much attention due to the homopolymerization and structural diversity of copolymerization products, and it is often used as a cross-linking agent in various functional materials.

Hydrogels are hydrophilic network polymers that absorb large amounts of water but are insoluble in water. The properties of high water content and insolubility in water make the materials to be widely used in biomedical fields such as drug delivery systems, biosensors, contact lenses, catheters and wound dressings [56]. Using DEGDVE as cross-linking agent and ^{60}Co as γ-radiation light source, hydrogels were prepared by radiation copolymerization of hydrophilic ethylene glycol monovinyl ether, hydrophobic BVE and acidic acrylic acid. The resulting hydrogels were pH- and temperature-sensitive functional materials [57–59].

Using DEGDVE as a cross-linking agent, 2-acrylamido-2-methylpropanesulfonic acid and sodium styrene sulfonate as monomers, a plugging agent with high temperature resistance and salt resistance was synthesized by emulsion polymerization, which has practical significance in solving the problem of water production in oil wells and plays an important role in improving oil recovery ratio.

A three-dimensional substance having a special space structure can be synthesized by cross-linking copolymerization with DEGDVE, and can be used as a modified substrate material. For example, cross-linking copolymerization of methyl vinyl sulfide with DEGDVE gave a solid polymer with a crown structure, which was modified and used as active matrices of solid superbases for ethynylation of acetone and vinylation of diethylene glycol or ethylene glycol [60].

References

[1] Hoelderich W, Goetz N, et al. Preparation of vinyl ethers. US 4891451, 1990.

[2] Dirk S, Holger M, Thomas R. Highly active catalyst systems for vinyl ether synthesis from acetylene and alcohols potassiumal coholate-grown ether in hydrocarbons, Journal of Organometallic Chemistry 1991, 414(2), C45–C50.

[3] Trofimov B A, Oparina L A, Lavlov V I, et al. Nucleophilic addition to acetylenes in superbasic catalytic systems. Part 7. Vinylation of low-molecular alcohols, Zhurnal Organicheskoi Khimii 1995, 31(5), 647–650.

[4] Vitkovskaya N M, Larionova E Y, Kobychev V B, et al. Methanol vinylation mechanism in the KOH/DMSO/CH3OH/C2H2 system, International Journal of Quantum Chemistry 2011, 111(11), 2519–2524.
[5] Kyoji A, Hideki H. Production of methyl vinyl ether. JP 4095040, 1990.
[6] Oliver D W H. Manufacture of vinyl ethers. GB 1035396, 1964.
[7] Trimm D, Cant N, Lei Y. Oxygenated fuel additives: The formation of methyl vinyl ether and 1,1-dimethoxyethane by the catalysed reaction of acetylene with methanol, Catalysis Today 2009, 145, 163–168.
[8] Taguchi Y, Oishi A, Shibuya I, et al. Method for the preparation of β-Lactam compound. US 6407228, 2002.
[9] Toshio E, Fujii T, Tsuji Y, et al. Method for producing copolymer of alkylvinylether and maleic anhydride, and copolymer of alkylvinylether and maleic anhydride. US 20020086964, 2002.
[10] Field N D, Lorenz D H. Process of preparing powdery copolymers of vinyl ethers and maleic anhydride using polymeric anti-clumping additives derived from vinyl alkoxy alkyl ethers. US 3,532,771, 1970.
[11] Mello S V, Arvanitidou E, Stranick M A, et al. Mode of action studies of a new desensitizing mouthwash containing 0.8% arginine, PVM/MA copolymer, pyrophosphates, and 0.05% sodium fluoride, Journal of Dentistry 2013, 41S, S12–S19.
[12] Ghosh A, Carran R S, Grosvenor A J, et al. Feather meal-based thermoplastics: methyl vinyl ether/maleic anhydride copolymer improves material properties, Fibers and Polymers 2016, 17(1), 9–14.
[13] Varshosaza J, Minaiyanb M, Dayyani L. Poly(methyl vinyl ether-co-maleic acid) for enhancement of solubility, oral bioavailability and anti-osteoporotic effects of raloxifene hydrochloride, European Journal of Pharmaceutical Sciences 2018, 112, 195–206.
[14] Francois P, Vaudaux P, Nurdin N, et al. Physical and biological effect of a surface coating procedure on polyurethane catheters, Biomaterials 1996, 17(7), 667–678.
[15] Reid G, Lam D, Policova Z, et al. Adhesion of two uropathogens to silicone and lubricious catheter influence of pH, urea and creatinine, Journal of Materials Science. Materials in Medicine 1993, 4(1), 17–22.
[16] Salman H H A, Azcarate I G. Nanoparticles comprising esters of poly (methyl vinyl ether-co-maleic anhydride) and uses thereof. US 9,351,940 B2, 2016.
[17] Dun J, Zhao D. Discussion and optimization of synthesis process of vinyl n-butyl ether, Gansu Chemical Industry 2004, 3, 26–29.
[18] Oparina L A, Shaikhudinova S I, Parshina L N, et al. Nucleophilic addition to acetylenes in superbasic catalytic systems: XIII. fluoride cesium containing systems, efficient catalysts for alkanols vinylation, Russian Journal of Organic Chemistry 2005, 41(5), 656–660.
[19] Zhang Y, Yang J, Xiong Y, et al. Polymerization method of n-butyl vinyl ether high polymer. CN 102627715 A, 2012.
[20] Reddy C K, Raju L Y, Srinivas P V S S, et al. Preparation and characterization of core-shell nanoparticles containing poly (chlorotrifluoroethylene-co-ethyl vinyl ether) as core, Journal of Applied Polymer Science 2011, 122(3), 1807–1814.
[21] Tanaka K, Ishida Y, Yanagi K, et al. Method of producing α, β-unsaturated ether. US 4,014,941, 1977.
[22] Mizuno K, Kimura Y, Otsuji Y. A convenient synthesis of aryl vinyl ethers by use of tetra n-butylammonium hydrogen sulfate, Synthesis 1979, 1979(9), 688–688.
[23] Okimoto Y, Sakaguchi S, Ishii Y. Development of a highly efficient catalytic method for synthesis of vinyl ethers, Journal of the American Chemical Society 2002, 124(8), 1590–1591.
[24] Lin J. Synthesis of isobutyl vinyl ether, Zhejiang Chemical Engineering 2006, 37(12), 1–4.
[25] Liu L. Method for synthesizing vinyl isobutyl ether by continuous method. CN 02138285.9, 2002.

[26] Meng Y. Continuous synthesis method of vinyl isobutyl ether. CN 200510040652.9, 2005.

[27] Li S F. An installation and method for preparation of isobutyl vinyl ether. CN 201210570442.0. 2014.

[28] Schildknecht C E, Zoss A O, Mckinley C. Vinyl alkyl ethers, Industrial and Engineering Chemistry 2002, 39(2), 180–186.

[29] Daniels W E. Emulsion copolymerization of isobutylvinyl ether and vinyl chloride. US 3,741,946, 1973.

[30] Chen F, Ye F, Chu G, Guo J, Huo L. Synthesis of acrylate modified vinyl chloride and vinyl isobutyl ether copolymers and their properties, Progress in Organic Coatings 2010, 67, 60–65.

[31] Liu S, Sun X J. Overview of production technology of vinyl chloride-isobutyl vinyl ether copolymer resin, Polyvinyl Chloride 2014, 42(4), 11–17.

[32] Zhou L, Yang Z R. Development of photocuring screen printing inks containing vinyl ether, Package Engineering 2006, 27(3), 8–10.

[33] Wang J F, Tian J, Ye P. Co-polymerization of isobutyl/4-hydroxybutyl vinyl ether and maleic anhydride, Designed Monomers and Polymers 2010, 13, 1–11.

[34] Endo T, Fujii T, Tsuji Y, et al. Method for producing copolymer of alkyl vinyl ether and maleic anhydride, and copolymer of alkyl vinyl ether and meleic anhydride. US 20020086964 A1, 2002.

[35] Liu S J, Li Z, Ji S J, et al. Synthesis of isobutyl vinyl ether-maleic anhydride and its application in preparation of chlorpyrifos microcapsules, Anhui Agricultural Science 2012, 40(9), 5317–5318.

[36] Oparina L A, Khil'ko M Y, Chernyshova N A, et al. Nucleophilic addition to acetylenes in superbasic catalytic systems: XIV. vinilation of diols in a system CsF-NaOH, Russian Journal of Organic Chemistry 2005, 41(5), 661–667.

[37] Shinya Y, Takaaki F, Shinji N, et al. Carbonate bond-containing hydroxyalkylallyl (or vinyl) ether, its composition, fluorocopolymer and its compositions. JP 9025319A, 1997.

[38] Wang G, Liu L, Xu G. Steel structure fireproof coating for preventing hydrocarbon fire and preparation method thereof. CN 1927962, 2007.

[39] Chang K, Vara F J. Radiation curable compositions containing multifunctional vinyl ether monomers and protective coatings formed therefrom. US 5106885, 1992.

[40] Peng C, Cationic Z W. UV curing of vinyl ether modified epoxy resin, China Coatings Industry 2002, 32(8), 17–19.

[41] Zhang J, Zhang Y, Bai C. Study on continues synthesis process of diethylene glycol vinyl ether, Journal of Jilin Institute of Chemical Technology 2012, 29(3), 20–23.

[42] Zhao F, Zhang Y, Zhang J, et al. Synthesis of diethylene glycol vinyl ether in bubble reactor, Journal of Jilin Institute of Chemical Technology 2013, 30(3), 6–9.

[43] Jiang W, Tian Y, Liu Y. The preparation of di-ethylene glycol divinyl ether, Science & Technology Vision 2014 (23), 183–184.

[44] Huang X Q, Li S F. Preparation of diethylene glycol vinyl ether by liquid phase cyclic reaction in tubular reactor, Chemical Industry and Engineering Progress 2018, 37(A1), 187–192.

[45] Lorenz R E. Separation of the divinyl ether of diethylene glycol or triethylene glycol from the monovinyl ether of the corresponding oliogethylene glycol. US 6,093,855, 2000.

[46] Maidhof T, Kronemayer H, Grosch G H, et al. Process for preparing divinyl ethers. US 8,946,487 B2, 2015.

[47] Tang M, Wang S. Preparation of UV-curable epoxy-acrylate photosensitive coating, Chemical Industry and Engineering Progress 2010, 29(6), 1116–1119.

[48] Liu L, Chen Y. Vinyl ether – a new type of reactive diluents for photocuring, Polymer Bulletin 1997 (4), 242–245.

[49] Nowak R M. Method of preparing the homopolymer of the monovinyl ether of diethylene glycol. US 3328468, 1967.

[50] Nurkeyeva Z S, Shaikhutdinov Y M, Seitov A Z, et al. Radiation polymerization of vinyl ethers of glycols and amino alcohols, Polymer Science U.S.S.R. 1987, 29(5), 1032–1038.
[51] Chu Q, Wolfgang K A, Förch R. Pulsed plasma polymerized di(ethylene glycol) monovinyl ether coatings for nonfouling surfaces, Chemistry of Materials 2006, 18(20), 4840–4844.
[52] Zhang Z, Menges B, Timmons R B, et al. Surface plasmon resonance studies of protein binding on plasma polymerized di(ethylene glycol) monovinyl ether films, Langmuir 2003, 19(11), 4765–4770.
[53] Sterling G B, Zimmerman R L. Rubbery copolymer of monovinyl ether of diethylene glycol/nu-vinyl pyrrolidinone and vulcanizate thereof. US 7829460 A, 1964.
[54] Pei D F, Chen S, Li S Q, et al. Fabrication and properties of poly(polyethylene glycol n-alkyl ether vinyl ether)s as polymeric phase change materials, Thermochimica Acta 2016, 633, 161–169.
[55] Liu W, Nichols P J, Smith N. Di(ethylene glycol) vinyl ether: a highly efficient deactivating reagent for olefin metathesis catalysts, Tetrahedron Letters 2009, 50(45), 6103–6105.
[56] Byrne M E, Park K, Peppas N A. Molecular imprinting within hydrogels, Advanced Drug Delivery Reviews 2002, 54(1), 149–161.
[57] Gümüşderelioğlu M, Topal I U. Vinyl ether/acrylic acid terpolymer hydrogels synthesized by γ-radiation: characterization, thermosensitivity and pH-sensitivity, Radiation Physics and Chemistry 2005, 73(5), 272–279.
[58] Kiremitçi A S, Ciftçi A, Özalp M, et al. Novel chlorhexidine releasing system developed from thermosensitive vinyl ether-based hydrogels, Journal of Biomedical Materials Research. Part B, Applied Biomaterials 2007, 83B(2), 609–614.
[59] Karakeçili A G, Satriano C, Gümüşderelioğlu M, et al. Thermoresponsive and bioactive poly(vinyl ether)-based hydrogels synthesized by radiation copolymerization and photochemical immobilization, Radiation Physics & Chemistry 2008, 77(2), 154–161.
[60] Trofimov B A, Morozova L V, Mikhaleva A I, et al. Modified copolymers of bifunctional vinyl ethers with methyl vinyl sulfide as active matrices of solid superbases, Russian Chemical Bulletin 2008, 57(10), 2111–2116.

Chapter 4
Derivatives from acetylene reacting with ketone

At present, the commonly used alkynol compounds include propargyl alcohol, methyl butynol, methylpentynol, 3-methyl-2-penten-4-yn-1-ol, 2,5-dimethyl-3-hexyne-2, 5-diol, 3,6-dimethyl-4-octyne-3, 6-diol, ethyl octynol and 2,4,7,9-tetramethyl-5-decyne-4, 7-diol. The physical properties of these alkynols are different but their structures are similar. In addition to alkynyl group, all of them contain polar hydroxyl group and nonpolar alkyl group. The structure determines that alkynol compounds have many excellent properties such as surface activity, wettability, defoaming property, dispersibility and corrosion inhibition.

4.1 Methylbutynol and isoprene

Methylbutynol, also known as 2-methyl-3-butyn-2-ol, α,α-dimethylpropargyl alcohol, 1,1-dimethyl-2-propynol, 1,1-dimethylpropargyl alcohol, 2-methyl-2-butynol, 2-methyl-3-butyne-2-ol, 3-methyl-1-butyn-3-ol, 3-hydroxy-3-methyl-1-butyne, 3-methylbutynol and 1,1-dimethylpropynol, is a derivative of acetylene and acetone. CAS Registry Number: 115-19-5. Molecular formula: C_5H_8O, molecular weight: 84.1164. The chemical structure is

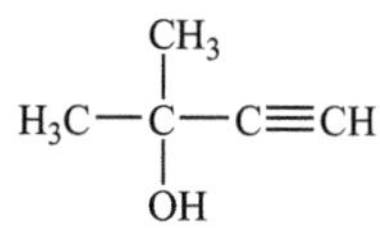

Isoprene, also known as isopentadiene, 2-methyl-1,3-butadiene and 2-methylbutadiene, is a derivative of methylbutynol. CAS Registry Number: 78-79-5. Molecular formula: C_5H_8, molecular weight: 68.1170. The chemical structure is

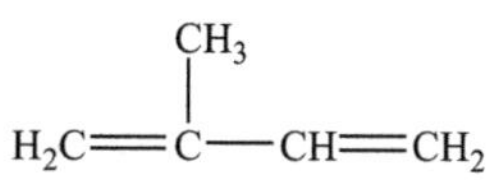

4.1.1 Properties of methylbutynol and isoprene

Methylbutynol is a colorless, transparent and aromatic liquid at room temperature. It is soluble in water and fatty acids, and miscible with acetone, benzene, carbon tetrachloride and petroleum ether but insoluble in ammonia. Methylbutynol and

https://doi.org/10.1515/9783110714999-004

water form an azeotrope having an azeotropic point of 91 °C in the methylbutynol composition of 74% or water composition of 26%. Table 4.1 lists the main physical properties of methylbutynol. Methylbutynol is chemically active and has the common reactions of alcohol groups and unsaturated triple bonds, such as addition, polymerization, esterification and oxidation reactions.

Table 4.1: Physical properties of methylbutynol.

Melting point/°C	3.0	Flash point (open cup)/°C	25
Boiling point/°C	104	Enthalpy of vaporization/ kJ · mol^{-1}	
Vapor pressure (20 °C)/mmHg	15	36 °C	49.5
Density (20 °C)/g · cm^{-3}	0.8614	64 °C	43.9
Refractive index (25 °C)	1.4184	80 °C	41.0

Isoprene is a colorless, volatile and irritating oily liquid at room temperature. It is almost insoluble in water but soluble in most organic solvents such as hydrocarbons, alcohols and ethers. Isoprene can form binary and ternary azeotropes with various organic substances. The physical properties of isoprene are shown in Table 4.2. Isoprene is chemically active and easy to polymerize.

Table 4.2: Physical properties of isoprene.

Melting point/°C	−145.95	Flash point (open cup)/°C	−48
Boiling point/°C	34.1	Enthalpy of fusion (−145.9 °C)/kJ · mol^{-1}	4.92
Density (20 °C)/g · cm^{-3}	0.6809	Enthalpy of vaporization/ kJ · mol^{-1}	
Heat capacity/J · mol^{-1} · K^{-1}		−4 °C	28.3
$C_{p,liquid}$ (25 °C)	152.5	26 °C	27.3
$C_{p,gas}$ (127 °C)	129.59	34 °C	25.8

4.1.2 Production processes of methylbutynol and isoprene

Methylbutynol is prepared by the reaction of acetylene with acetone. The commercial processes for isoprene production include extraction method (recovery from C5 fraction), dehydrogenation of isopentane or isopentene, reaction of isobutylene with formaldehyde and the acetylene–acetone route. Here, only the acetylene–acetone route for the production of isoprene is described. Acetone underwent an ethynylation reaction with acetylene to yield methylbutynol, followed by partial hydrogenation to 2-methyl-3-buten-2-ol and finally dehydration to the desired product. The reaction equations are as follows:

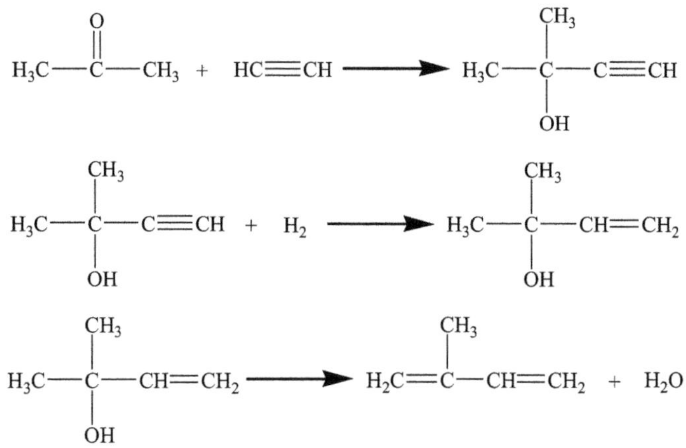

4.1.2.1 Ethynylation

Three parts of the strong basic anion exchange resin Amberlite IRA400 were suspended in 30 parts of 7% sodium hydroxide solution to change them to OH type. After washing to neutral with water, the resin was suspended in methanol, and after 15–60 min, methanol was discharged. The treatment with methanol was repeated as described above until the water content of the effluent methanol was less than 0.1%, and the water content of the ion exchange resin was less than 0.2%. The tubular reactor was filled with the ion exchange resin, and then acetone saturated with acetylene and liquid ammonia was introduced into the reactor at 2.5 MPa and 40 °C. Excess acetylene in reactants was used to reduce the side reaction of acetone. After ethynylation, the reaction solution was subjected to a flash evaporation at 50 °C. Ammonia and the unreacted acetylene were recovered and recycled. The unreacted acetone was recovered by distillation and recycled. The remaining solution was distilled to obtain methylbutynol. The conversion of acetone was 90% and the selectivity to methylbutynol was 95%. After continuous operation for 25 days, the ion exchange resin was required to be regenerated.

4.1.2.2 Selective hydrogenation

The methylbutynol and hydrogen were introduced into the hydrogenation reactor filled with Pd/$CaCO_3$ catalyst. The hydrogenation was carried out at 50–80 °C and under 0.5 MPa. The main by-product is methylbutanol. In order to suppress the formation of by-products, it is necessary to add a small amount of zinc salt as an inhibitor. The conversion of the reaction was 100%, and the selectivity to methylbutenol was between 98% and 99%. The reaction mixture was separated in a gas–liquid separator. The unreacted hydrogen was recycled. The liquid was evaporated and finally sent to a dehydration reactor.

4.1.2.3 Dehydration

The azeotrope of methylbutenol and water was pumped to a fixed-bed dehydration reactor filled with alumina catalyst. The dehydration of methylbutenol was carried out at 260–300 °C and under atmospheric pressure. The conversion of methylbutenol was from 88% to 98%, and the selectivity to isoprene was more than 98%. The dehydrated product was washed with water in a washing column to remove a small quantity of water-soluble substances, and then sent to two distillation towers in series for removal of the light fraction and the heavy fraction, respectively. Isoprene of polymerization grade was obtained at the top of the heavy fraction removing tower.

The process flow diagram for the manufacture of methylbutynol and isoprene from acetylene and acetone is shown in Figure 4.1.

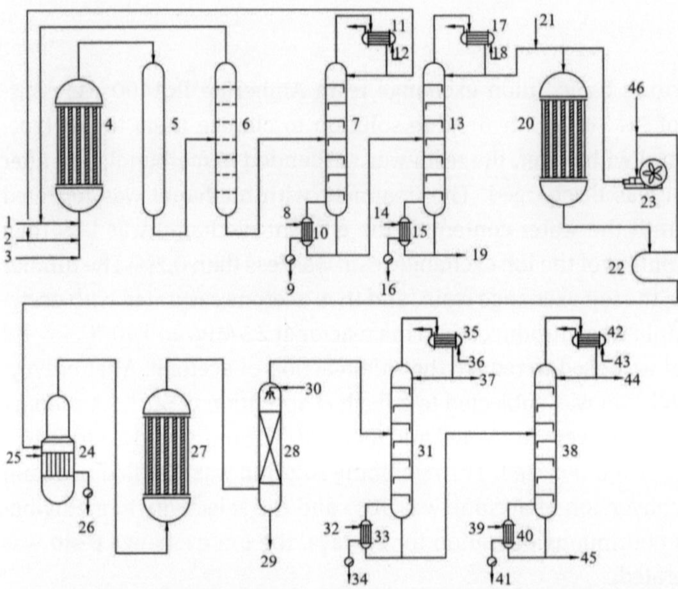

1. Acetylene feed; 2. Acetone feed; 3. Catalyst; 4. Ethynylation reactor; 5. Reaction termination tower; 6. Flash tower; 7. Acetone recovery tower; 8, 14, 25, 32, 39. Steam; 9, 16, 26, 34, 41. Condensate drain; 10, 15, 33, 40. Reboiler; 11, 17, 35, 42. Condenser; 12, 18, 36, 43. Cooling water; 13. Heavy fraction removing tower; 19. Heavy fraction; 20. Hydrogenation reactor; 21. Inhibitor feed; 22. Gas-liquid separator; 23. Compressor; 24. Evaporator; 37. Dehydration reactor; 28. Washing tower; 29. Drain; 30. Water feed; 31. Light fraction removing tower; 37. light fraction; 38. Heavy fraction removing tower; 44. Ispprene; 45. Heavy fraction; 46. Hydrogen

Figure 4.1: Process flow diagram for the manufacture of methylbutynol and isoprene from acetylene and acetone.

4.1.3 Uses of methylbutynol and isoprene

Methylbutynol is used as an intermediate in the pharmaceutical industry for the synthesis of the anticancer drug citricin. It is used as a raw material for the synthesis of pyrethroids in the pesticide industry, and also used as a raw material in the synthesis of terpene perfumes. Because methylbutynol has good corrosion inhibition and synergistic effect, it can effectively prevent hydrogen embrittlement of metal. As a basic component of high-efficiency acidification corrosion inhibitor, it is widely used in oil and gas industry.

Methylbutynol can be applied as an acetylene carrier for safe and economical transport of large acetylene quantities [1]. This usage can be accomplished by three steps. Firstly, acetylene is stored in the form of methylbutynol by reacting the former with acetone. Secondly, the resulting methylbutynol is transported from the origin of acetylene to the place of consumption. Thirdly, acetylene and acetone can be obtained by cleavage of methylbutynol and separated to yield pure products for further applications. Figure 4.2 shows the process flow diagram for the cleavage of methylbutynol in a continuous manner. Methylbutynol and catalyst are fed into the cleavage reactor that is operated at 80 °C. The reaction mixture is removed from the reactor and separated by a distillation tower. Acetylene is recovered as a pure product. Unreacted methylbutynol exits from the bottom of the distillation tower and returns to the reactor to continue the reaction. The liquid condensed at the top of the tower still contains acetylene and is separated in a gas–liquid separator. The liquid from the bottom of the gas–liquid separator is further separated by another distillation tower to recover pure acetone.

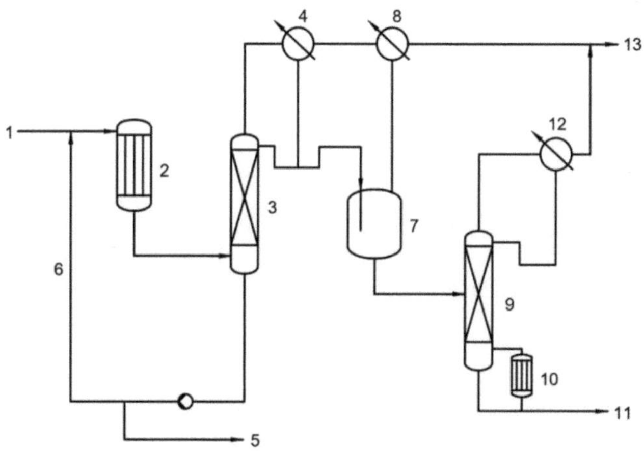

1. Methylbutynol and catalyst; 2. Cleavage reactor; 3, 9. Distillation tower; 4, 8, 12. Condenser; 5. Residue; 6. Methylbutynol recycle; 7. Gas-liquid separator; 10. Reboiler; 11. Acetone; 13. Acetylene.

Figure 4.2: Process flow diagram for the cleavage of methylbutynol.

Isoprene can be applied to produce polyisoprene rubber by solution polymerization, or can be copolymerized with isobutylene to form butyl rubber, and can also be used for the production of synthetic fragrances and medicines.

4.2 Methylpentynol

Methylpentynol, also known as 3-methyl-1-pentyn-3-ol, meparfynol, α-ethyl-α-methylpropargyl alcohol, 3-methyl-1-pentyne-3-ol and 3-hydroxy-3-methyl-1-pentyne, is a derivative of acetylene and butanone. CAS Registry Number: 77-75-8. Molecular formula: $C_6H_{10}O$, molecular weight: 98.1430. The chemical structure is

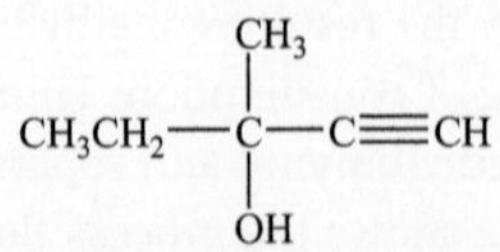

4.2.1 Properties of methylpentynol

Methylpentynol is a colorless liquid with low viscosity at room temperature. It has a spicy odor and a burnt smell. The solubility of methylpentynol in water is 12.8% at 25 °C. It is miscible with acetone, benzene, carbon tetrachloride, cyclohexanone, ethylene glycol ether, diethylene glycol, ethyl acetate, kerosene, methyl ethyl ketone, ethanolamine, hoof oil, diethyl ether, petroleum ether, soybean oil, dry-cleaning solvent gasoline and the like. Its density is 0.8688 g cm^{-3} at 25 °C. The melting point is −30.6 °C, the boiling point is 121–122 °C and the flash point is 38 °C. The refractive index is 1.4318 at 20 °C. The vapor pressure is 6.5 mmHg at 20 °C. It is flammable and toxic.

4.2.2 Production process of methylpentynol

Methylpentynol is prepared by ethynylation of butanone with acetylene:

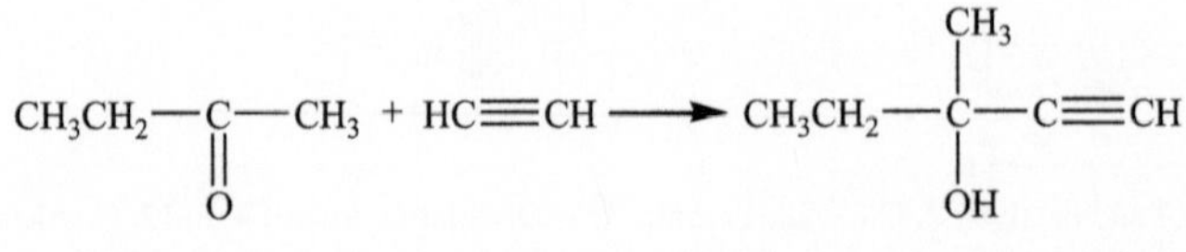

At present, there are three main processes for production of methylpentynol in the industry: ethynylation catalyzed by solid potassium hydroxide, ethynylation catalyzed by liquid ammonia and potassium hydroxide and ethynylation catalyzed by potassium alkoxide.

4.2.2.1 Ethynylation catalyzed by solid potassium hydroxide

In this process, acetylene is introduced into a solution of an organic solvent (e.g., aromatic hydrocarbon, dichloromethane, diethyl ether and tetrahydrofuran) and butanone in which the powdery solid potassium hydroxide is suspended, and the ethynylation reaction is carried out at near room temperature and under atmospheric pressure. The greatest advantage of this process is that the reaction can be carried out under atmospheric pressure, so that the requirements for equipment are not so harsh, the operation is also very convenient and safe and the equipment investment is low. However, this process also has some shortcomings: First, the powdery potassium hydroxide catalyst will be dissolved in water to form an aqueous potassium hydroxide solution after completion of the reaction, making it difficult to recycle. Since the consumption of solid potassium hydroxide in this method is 3 to 4 times of the consumption of the raw material butanone in terms of moles, the consumption of solid potassium hydroxide is very large and the blockage phenomenon is often caused in the reaction. The treatment of a large amount of alkali liquor is also a difficult problem. Second, due to the use of potassium hydroxide powder, the reaction mixture is particularly viscous and difficult to be fully stirred uniformly in a typical stirred tank. In order to prevent clogging and achieve good mass and heat transfer conditions, a special stirring device or a large amount of solvent is required. Third, the batch operation has a long operation cycle for more than 20 h, and the yield of methylpentynol is as low as less than 70%.

4.2.2.2 Catalytic ethynylation in liquid ammonia–KOH system

Acetylene is dissolved in liquid ammonia, and butanone and potassium hydroxide catalyst are added. In this process, liquid ammonia not only acts as a solvent but also acts as a cocatalyst. Potassium hydroxide is in the form of an aqueous solution with a mass concentration of 50%, which is used in a very small amount and easy to be recycled. The ethynylation reaction is carried out at a temperature of 30 °C and under a pressure of 1.6 MPa for 1 h. The mass ratio of liquid ammonia to butanone is 2.5:1, and the molar ratio of KOH to butanone is 1:359. The yield of methylpentynol is more than 90% based on butanone [2]. Compared with the ethynylation catalyzed by solid potassium hydroxide, the consumption of KOH is greatly reduced. The raw materials and products are all in the liquid phase, and the reaction is homogeneous. It is easy to transfer heat and mass, and easy to control the reaction temperature. It is also easy to achieve continuous production and expand production scale. However, the disadvantage of this process is that the liquid ammonia is highly volatile and requires a pressurization operation, which requires high standard equipment.

4.2.2.3 Catalytic ethynylation by potassium alkoxide

Potassium alkoxide is used as a catalyst instead of potassium hydroxide. A good catalytic effect can be achieved by using equimolar potassium alkoxide and butanone. The potassium alkoxide can be prepared using potassium hydroxide and an alcohol under atmospheric pressure. Once the synthesis reaction of methylpentynol is completed, the potassium alkoxide can be further decomposed into an alcohol and potassium hydroxide, so that potassium hydroxide can be recycled.

KOH aqueous solution and isobutanol were added to the catalyst preparation kettle connected to a fractional distillation column. Steam was introduced into the jacket of the kettle for heating. During the reaction, the free water and water generated from reaction in the reaction system were removed. Once the reaction was completed, the potassium alkoxide was dispersed in xylene to obtain potassium alkoxide suspension. The suspension was pumped into the ethynylation reaction vessel. First, it was cooled to room temperature with circulating cooling water, and then cooled to near 0 °C with chilled brine. Subsequently, acetylene and butanone were introduced into the vessel. The reaction was carried out at 30 °C for 3 h. Then water was added in a dropwise manner for hydrolysis. The mixture was allowed to stand for separating into two layers. The oil layer was collected and washed with water, and subjected to fractional distillation for methylpentynol product. The yield of methylpentynol was more than 90% based on butanone.

Catalytic ethynylation by potassium alkoxide has the advantages of operation at atmospheric pressure, less equipment investment, simple process and high yield, and versatility. One set of equipment can produce various alkynol according to production requirements.

In summary, although the solid KOH catalysis process has advantages including less one-time investment and simple operation, it is facing the situation of being eliminated because of its poor process reliability and serious environmental pollution. Catalytic ethynylation in liquid ammonia–KOH system has many advantages but the investment is huge, and a considerable production scale is required to produce better benefits. At present, the batch operation mode is more favorable for preparation of fine chemical product methylpentynol. Catalytic ethynylation by potassium alkoxide is a promising process for the production of methylpentynol, which combines the advantages of the above two processes. The key of this process is the preparation of potassium alkoxide catalyst, and if the time and energy consumed in the preparation of potassium alkoxide are too much, the value of this method will be greatly reduced.

4.2.3 Uses of methylpentynol

Methylpentynol was originally used only as a sedative. At that time, the research on it was not deep enough, and the high cost of production also limited its application. With the continuous discovery of the characteristics of methylpentynol and the continuous improvement of its synthesis technology, the use of methylpentynol is more and more extensive. For example, it can be used as a stabilizer for chlorinated solvents, a viscosity reducer and viscosity stabilizer and an intermediate for organic synthesis. In particular, methylpentynol is used as a corrosion inhibitor in the acidification fluid of oil and gas wells.

Acidification treatment is an effective stimulation measure to increase the oil and gas recovery in oil and gas wells by treating carbonate or sandstone rock layers with hydrochloric acid. The concentration of hydrochloric acid used is usually 10–15%, and recently it has been increased to 28% but the acidification operation also brings serious corrosion to downhole facilities. Moreover, the metal iron ions dissolved by acid may also cause damage to the stratum. Therefore, in the acidification of oil and gas wells, in order to solve the corrosion problem of the high-temperature acidification liquid on the tubing equipment, it is necessary to add a corrosion inhibitor to the hydrochloric acid medium.

Organic compounds that can be strongly adsorbed on the metal surface are widely used as corrosion inhibitors in acidic media. The molecule of alkynol contains polar OH group, strong force carbon–carbon triple bond group and nonpolar alkyl group, so it can be used as an adsorptive organic corrosion inhibitor. Methylpentynol is an excellent corrosion inhibitor under the conditions of high temperature, high pressure and concentrated acid, which has stable performance and good synergistic effect with other corrosion inhibitors, and is the key component of corrosion inhibitor widely used in acidification fluid of high-temperature deep well in recent years. Methylpentynol has better corrosion inhibition performance than methylbutynol and dimethyl hexynediol. A corrosion inhibitor formulation comprising 0.1% methylpentynol, 0.35% phosphoric acid, 0.25% triethanolamine and 0.15% Sb_2O_3 was developed. It reduced the corrosion rate of 15% hydrochloric acid to N80 steel from 1,145 to 4.5 g/(m^2 · h) at a temperature of 80 °C and exhibited a corrosion inhibition rate as high as 99.6% by using its dosage of 0.85% based on the mass of hydrochloric acid solution [3].

4.3 *tert*-Pentanol

tert-Pentanol, also known as 2-methyl-2-butanol, *tert*-pentyl alcohol, *tert*-amyl alcohol, dimethyl ethyl carbinol, ethyl dimethyl carbinol and 1,1-dimethyl-1-propanol, is a derivative of methylbutynol. CAS Registry Number: 75-85-4. Molecular formula: $C_5H_{12}O$, molecular weight: 88.1482. The chemical structure is

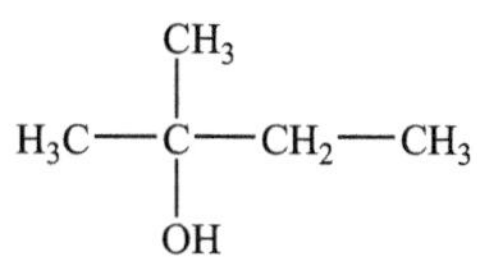

4.3.1 Properties of *tert*-pentanol

tert-Pentanol is a colorless, transparent and volatile liquid at room temperature with a special pungent odor. It is partially soluble in water. The solubility in water is 14% (mass fraction) at 30 °C, and it is miscible with ethanol, ether, benzene, chloroform, glycerin and the like. *tert*-Pentanol and water form an azeotrope at composition of 72.5% *tert*-pentanol and 27.5% water, having an azeotropic point of 87.35 °C. It can also form azeotropes with various organic solvents such as benzene, toluene, cyclohexane, methylcyclohexane, carbon tetrachloride, heptane, octane, propyl acetate and the like. The main physical properties of *tert*-pentanol are listed in Table 4.3.

Table 4.3: Physical properties of *tert*-pentanol.

Melting point/°C	−9.0	Flash point (open cup)/°C	21
Boiling point/°C	101.9	Enthalpy of vaporization (375.4 K)/kJ · mol^{-1}	39.04
Vapor pressure/kPa		Enthalpy of fusion (264.7 K)/kJ/ · mol^{-1}	2.0
17.2 °C	1.33	Heat capacity (25 °C)/J · mol^{-1} · K^{-1}	247.15
38.8 °C	5.33	Critical temperature/°C	272
46.0 °C	8.00	Critical pressure/MPa	3.71
55.3 °C	13. 3	Surface tension (20 °C)/(N/m)	22.77×10^{-3}
69.7 °C	26.6	Dipole moment/($\times 10^{-30}$ C · m)	1.7
85.7 °C	53.3	Dielectric constant (25 °C)	5.82
Density (20 °C)/g · cm^{-3}	0.8090	Viscosity (25 °C)/mP · s	3.70
Refractive index (20 °C)	1.4052	Explosion limit in air	1.2–8.0%

When *tert*-pentanol is exposed to air for a long time, it will be oxidized to high-boiling substances and the color turns yellow. Therefore, it is necessary to purge the container with a dry inert gas such as nitrogen to remove air and to maintain a protective atmosphere throughout the storage period. Since *tert*-pentanol is an organic solvent with excellent dissolving performance, it cannot be stored in a container of plastic and rubber. However, the storage and transportation container for *tert*-pentanol can be made of carbon steel, stainless steel or aluminum.

4.3.2 Process for manufacture of *tert*-pentanol

Industrial production methods of *tert*-pentanol include acetylene–acetone route and isopentene hydration route. The acetylene–acetone route uses acetylene from natural gas or calcium carbide as a raw material, while the isopentene hydration route uses isopentene from petroleum as a raw material. In China, *tert*-pentanol is produced by acetylene–acetone route.

Acetylene reacts with acetone to form methylbutynol. Hydrogenation of methylbutynol yields *tert*-pentanol:

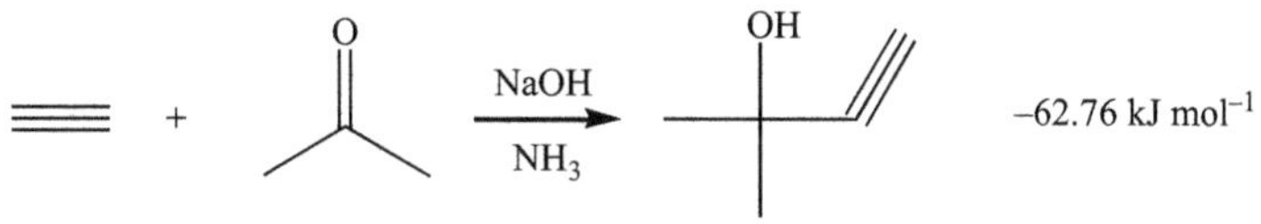

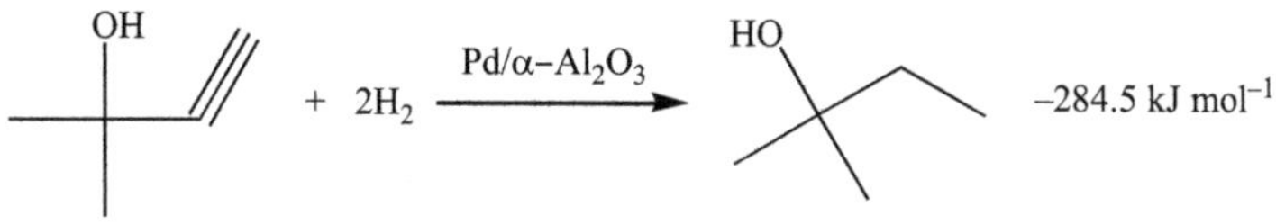

Acetylene and ammonia were mixed in a certain ratio, and the mixture was pressurized to a certain pressure by a compressor and condensed into a liquid by a condenser. Acetone is pumped into the reactor simultaneously with the condensate of the gas mixture according to a certain proportion. The catalyst sodium hydroxide was added to the reactor from the top and was in countercurrent contact with the reactants. After reaction, the reaction mixture was neutralized by HCl. The unreacted acetylene and ammonia were evaporated by a flash evaporator and recycled.

Acetone was recovered from the reaction solution after flash evaporation by an acetone recovery tower for reuse. The reaction solution was then subjected to fractional distillation for obtaining an azeotrope of methylbutynol and water.

The methylbutynol–water azeotrope was introduced into the first-stage hydrogenation reactor, and flowed down with hydrogen in the same direction through the Pd/α-Al_2O_3 catalyst layer. The outlet reaction mixture from the reactor was subjected to cooling and gas–liquid separation. Hydrogen was recovered, and the liquid was sent to the second-stage hydrogenation reactor. After the second-stage hydrogenation, the reaction mixture was cooled and separated from hydrogen. Water in the reaction liquid was removed by calcium chloride. Finally, pure *tert*-pentanol was obtained by fractional distillation.

The process using calcium chloride for water removal has some disadvantages including long process, large amount of wastewater, low yield of *tert*-pentanol and much consumption of calcium chloride. Afterward it has been improved by azeotropic distillation using cyclohexane as an entrainer [4]. The advantages are as follows: (1) calcium chloride is eliminated, and the cost of calcium chloride is saved; (2) the yield of *tert*-pentanol is increased by about 2.5%, and the cost is reduced; and (3) the environmental pollution caused by calcium chloride wastewater is eliminated.

Technical parameters of operation for *tert*-pentanol are as follows:

(1) Ethynylation
Volume ratio of acetylene/ammonia in the gas mixture: 1:3
Pressure: 2.0–2.2 MPa
Temperature: 35–40 °C
Residence time: 1.0–1.3 h
The yield of methylbutynol: ≥94%

(2) Hydrogenation
Hydrogen pressure: 0.65–0.70 MPa

Temperature: 35–80 °C
Space velocity: 1.5–2.0 h^{-1}
Volume ratio of gas to liquid in feed: 630 to 780:1
Catalyst life: ≥650 h
The conversion of hydrogenation: 100%
The yield of hydrogenation: ≥95%

The deactivated waste catalyst is regenerated by burning off the polymer with air at a temperature of 450–500 °C and then reduced by formaldehyde or hydrazine hydrate solution.

4.3.3 Uses of *tert*-pentanol

tert-Pentanol is mainly used as a solvent and an intermediate for some pharmaceuticals and fine chemicals. It is used as an intermediate for the synthesis of pesticides such as triazolone, triadimenol and vinyl azoles. *tert*-Pentanol can also be used for production of indane musk which is a perfume. Furthermore, *tert*-pentanol is a raw material for the manufacture of *tert*-amyl peroxypivalate, which is an initiator for the polymerization or copolymerization of ethylene, styrene, butadiene, acrylate and methacrylate. Moreover, *tert*-pentanol is used as a stabilizer for trichloroethylene, tetrachloroethylene, 1,1,1-trichloroethane, 1,2-dichloropropane, and so on. In these solvents, *tert*-pentanol has the effect of inhibiting their decomposition, volatilization and corrosion.

4.4 Dimethylhexynediol and dimethylhexanediol

Dimethylhexynediol, also known as 2,5-dimethyl-3-hexyne-2,5-diol, 2,5-dimethylhexyne-2,5-diol and tetramethylbutynediol, is a derivative of acetylene and acetone. CAS Registry Number: 142-30-3. Molecular formula: $C_8H_{14}O_2$, molecular weight: 142.1956. The chemical structure is

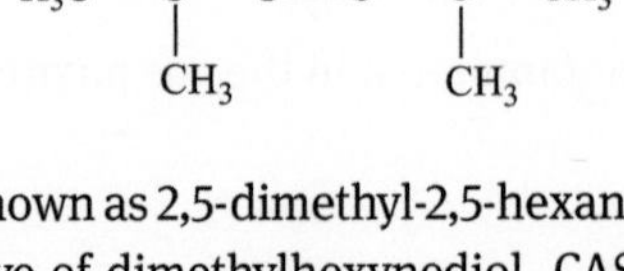

Dimethylhexanediol, also known as 2,5-dimethyl-2,5-hexanediol and 1,1,4,4-tetramethyl-1,4-butanediol, is a derivative of dimethylhexynediol. CAS Registry Number: 110-03-2. Molecular formula: $C_8H_{18}O_2$, molecular weight: 146.2273. The chemical structure is

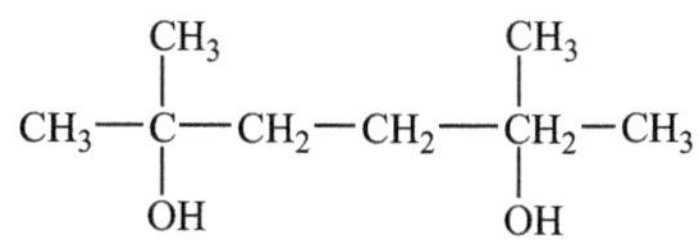

4.4.1 Properties of dimethylhexynediol and dimethylhexanediol

Dimethylhexynediol is a white crystal with a melting point of 94.5 °C and a boiling point of 205.5 °C. The vapor pressure is 0.9 kPa at 122 °C. Dimethylhexanediol is also a white crystal with a melting point of 89 °C and a boiling point of 214–215 °C. The density of dimethylhexanediol is 0.898 at 20 °C. Both dimethylhexynediol and dimethylhexanediol are soluble in water, acetone and alcohol, but they are insoluble in benzene, carbon tetrachloride and kerosene.

4.4.2 Production process of dimethylhexynediol and dimethylhexanediol

Dimethylhexynediol and dimethylhexanediol are mainly produced by ethynylation of acetone (acetylene–acetone process) or by condensation of methylbutynol and acetone (alkynol–acetone process). In China, the production of dimethyl hexynediol by the acetylene–acetone process has been started from the late 1960s to the early 1970s.

4.4.2.1 Manufacture of dimethylhexynediol by ethynylation of acetone

In the presence of potassium hydroxide catalyst, acetylene and acetone underwent a condensation reaction, then the reaction mixture was hydrolyzed and neutralized to obtain dimethylhexynediol:

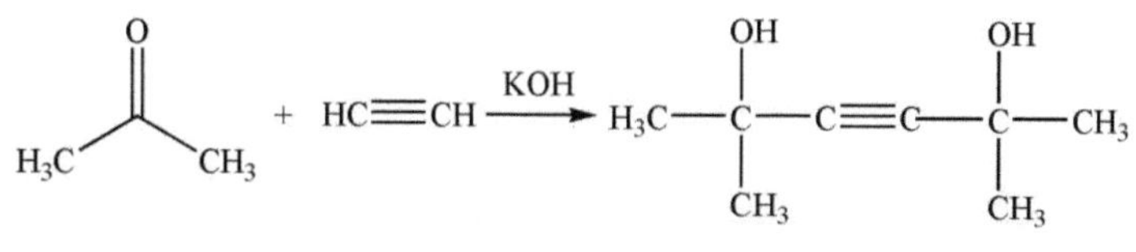

Finely divided and water-free potassium hydroxide powder was used as catalyst and suspended in organic solvents such as benzene or methyl *tert*-butyl ether. Acetone and acetylene were used in a molar ratio of 1:1 to 3:1 while KOH and acetone in a molar ratio of 1:1 to 1.6:1. The reaction was carried out at a temperature between 20 and 35 °C for 4 h. Subsequently, water was added to hydrolyze the reaction mixture for 1 h. The organic phase was separated and neutralized with glacial acetic acid. The conversion of acetone was 92% and the yield of dimethylhexynediol was 72%.

There are several disadvantages in the above process. First, the catalyst is used in large quantities and cannot be recycled, resulting in the generation of large amounts of wastewater. Second, the process mostly produces an appreciable proportion of by-product such as methylbutynol. Third, the reaction mixture becomes so viscous due to the formation of adducts of KOH and dimethylhexynediol that stirring is difficult toward the end of the reaction.

The process has been much improved by using potassium isobutoxide in place of potassium hydroxide. Usually, a suspension of solid potassium isobutoxide in xylene is applied. The conversion of acetone and the yield of dimethylhexynediol are enhanced to 98% and 86%, respectively. In addition, methylbutynol is obtained as a by-product in a 4% yield [5].

4.4.2.2 Manufacture of dimethylhexynediol by condensation of methylbutynol and acetone

Using methylbutynol and acetone as raw materials and xylene as a solvent, condensation reaction of methylbutynol with acetone is carried out in the presence of potassium isobutoxide catalyst to synthesize dimethylhexynediol. The chemical reaction equation is as follows:

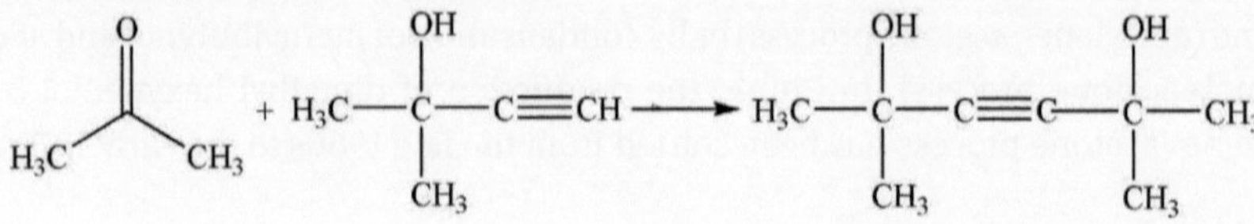

The synthesis process is carried out in two steps. First, isobutanol reacts with potassium hydroxide aqueous solution to form potassium isobutoxide, and then methylbutynol reacts with acetone to form dimethylhexynediol in the presence of potassium isobutoxide.

The preparation of potassium isobutoxide is a key step of the process. It is carried out in a double-walled stirred vessel connected to a fractional distillation column. Isobutanol and xylene are added to the vessel. An aqueous solution of potassium hydroxide having a concentration of 33% by weight is introduced into the upper section of the column. The temperature at the top of the column is about 91 °C, and the temperature at the bottom is 140 °C. The azeotrope of water and isobutanol is removed from the top of the column, condensed into liquid by cooling water and separated into an organic layer and a water layer. A molar ratio of isobutanol and potassium hydroxide in the range from 4.5:1 to 9:1 is applied in order to complete the reaction. A suspension of potassium isobutoxide in xylene is obtained in the stirred vessel. The conversion of potassium hydroxide is more than 99%.

The suspension of potassium isobutoxide is cooled down to 0 °C, and then equimolar acetone and methylbutynol were added. The suspended solids disappeared, and the entire reaction system became a homogeneous solution. The condensation reaction is carried out at 25–35 °C for 3 h. After the reaction is completed, water is added for hydrolyzation for 2 h. The reaction mixture is allowed to stand for two layers, and an organic phase and an aqueous phase are separated. The aqueous phase contains potassium hydroxide and can be recycled for the preparation of potassium isobutoxide. The organic phase is neutralized with glacial acetic acid. Following partly removal of the solvent and cooling to room temperature, the dimethylhexynediol is obtained by crystallization. The overall yield of dimethylhexynediol is greater than 89%. Figure 4.3 shows the process flow diagram for production of dimethylhexynediol by methylbutynol–acetone route.

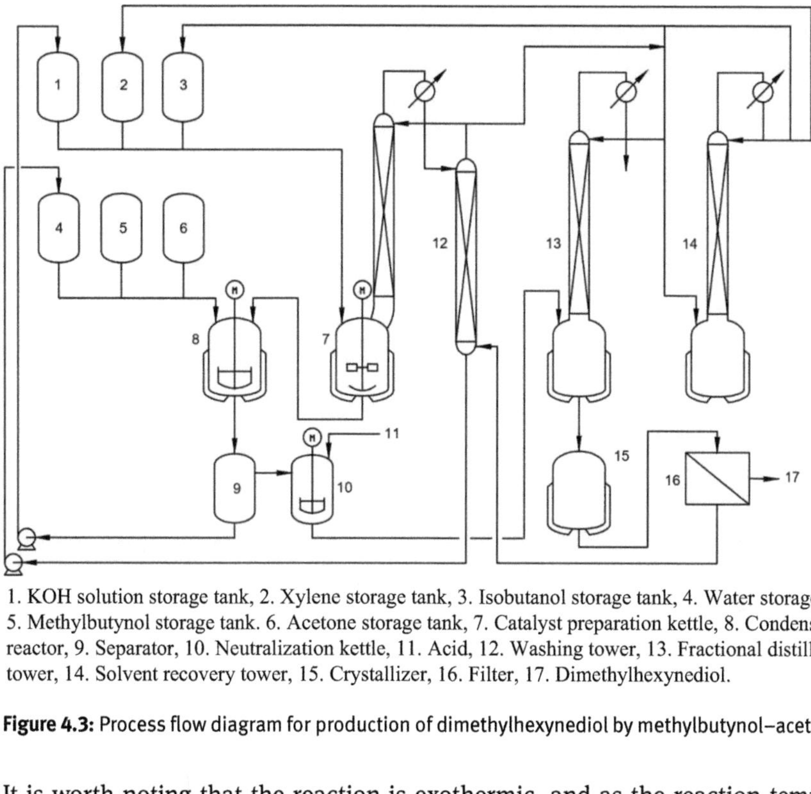

1. KOH solution storage tank, 2. Xylene storage tank, 3. Isobutanol storage tank, 4. Water storage tank, 5. Methylbutynol storage tank. 6. Acetone storage tank, 7. Catalyst preparation kettle, 8. Condensation reactor, 9. Separator, 10. Neutralization kettle, 11. Acid, 12. Washing tower, 13. Fractional distillation tower, 14. Solvent recovery tower, 15. Crystallizer, 16. Filter, 17. Dimethylhexynediol.

Figure 4.3: Process flow diagram for production of dimethylhexynediol by methylbutynol–acetone route.

It is worth noting that the reaction is exothermic, and as the reaction temperature increases, the yield of dimethylhexynediol decreases (Table 4.4).

Table 4.4: Effect of temperature on the yield of dimethylhexynediol.

Starting temperature (°C)	Reaction temperature (°C)	Yield (%)
35	50	80.65
25	38	87.36
15	35	88.60
10	30	89.00
0	30	89.63

In China, the alkynol–acetone process for production of dimethylhexynediol was developed by the Southwest Research Institute of Chemical Industry [6]. Its advantages include simple structure of the reaction equipment, convenient operation, significant reduction in production costs and less waste liquid discharge.

4.4.2.3 Production of dimethylhexanediol by hydrogenation of dimethylhexynediol

The hydrogenation of dimethylhexynediol comprises two consecutive reactions. The first reaction is the partial saturation of the acetylenic bond to a double bond, leading to the generation of dimethylhexenediol. Subsequently, the dimethylhexenediol is further hydrogenated to form dimethylhexanediol:

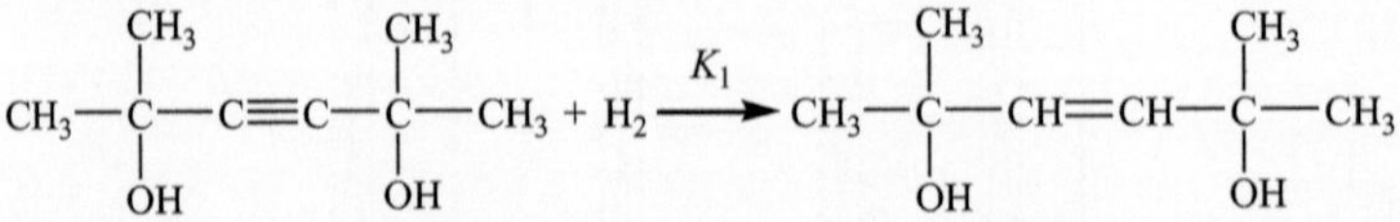

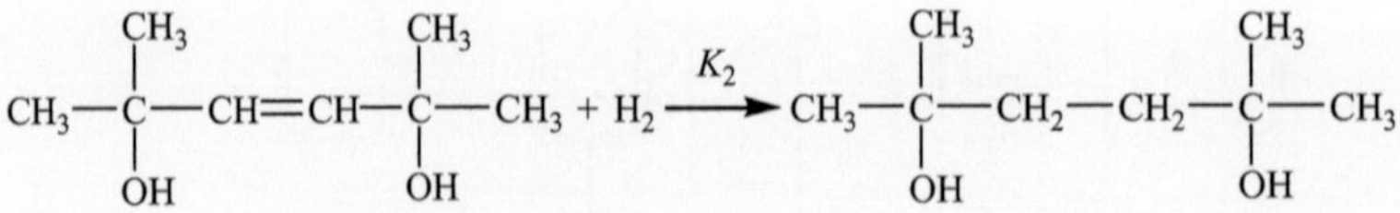

The reaction equilibrium constants K_1 and K_2 have been correlated as a function of temperature in the range from 300 to 345 K [7]:

$$\ln K_1 = 7.9397 - \frac{2{,}456.8}{T}$$

$$\ln K_2 = 4.9227 - \frac{1{,}529.8}{T}$$

The hydrogenation process of dimethylhexynediol can be conducted in a batch or continuous operation. For a batch operation, dimethylhexynediol and solvent are charged to a suitable hydrogenation vessel containing catalyst such as palladium supported on activated carbon. Suitable solvents are benzene, ethylbenzene and water. Hydrogen is then introduced into the vessel under the desired pressure. The reaction proceeds in the temperature range from 30 to 70 °C. After completion of the first phase of the reaction wherein the acetylenic bond is partially saturated as signified by adsorption of about one-half the hydrogen theoretically required, external heat is supplied to maintain the desired temperature and sustain the reaction. When the hydrogen pressure has dropped to the level indicating absorption of the amount of hydrogen theoretically required to completely saturate the acetylenic bond, the temperature is raised to a substantially higher temperature for a short period to ensure complete reaction. While the vessel is still hot, the contents are removed and filtered to separate the catalyst which can be reused without reactivation or replacement for another hydrogenation because of the unique action of the solvent. The recovered liquids are then treated by conventional methods to separate the saturated diol from the solvent. For example, the liquid reaction mixture is cooled to about 25 °C and the solid product that precipitates is separated by filtration from the liquid solvent and dried to recover the highly pure dimethylhexanediol.

In industry, continuous hydrogenation of dimethylhexynediol is preferred. For continuous operation, a trickle-bed reactor filled with spherical particles of supported catalyst (e.g., 0.25% Pd on aluminum oxide) is applied. Dimethylhexynediol solution (25–30%) is pumped quantitatively into the top of the trickle-bed reactor, and hydrogen is simultaneously introduced into the top and flows down through the catalyst bed. The outer jacket of the reactor is circulated with constant temperature water to control a certain temperature. After leaving the reactor, the output is depressurized into a vessel. The yield of dimethylhexanediol reaches 98.5% under optimized conditions.

The hydrogenation process is conducted under relatively moderate temperatures. Generally, the temperature may range from about 30 °C to about 80 °C. If hydrogenation is carried out at too low a temperature, the time required for complete saturation is generally too long for commercial operation; and if the hydrogenation is conducted at too high a temperature, the production of the undesired hydrogenolysis products is favored. There are two side reactions in the reaction process: one is the dehydration reaction, which generates dialkene, and then the dialkene can be hydrogenated to alkane:

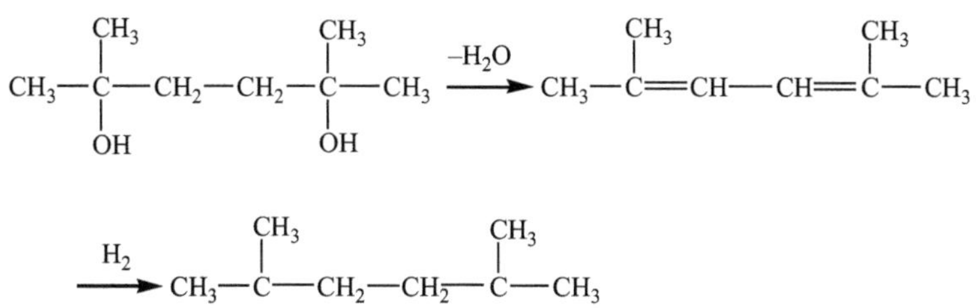

Another side reaction is the decomposition of dimethylhexynediol:

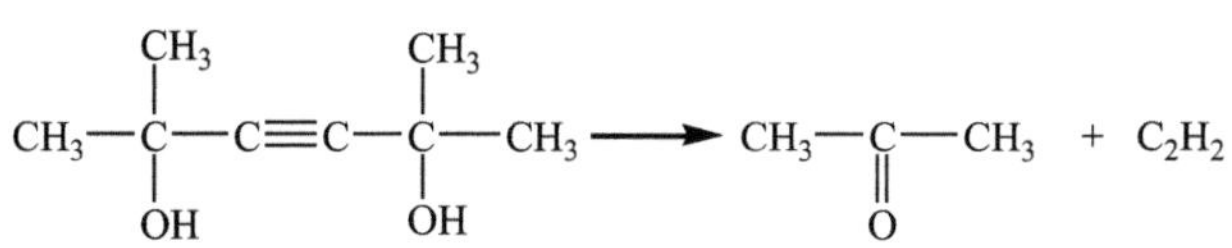

These side reactions occur at higher temperatures, leading to a decrease in the yield of dimethylhexanediol. The optimized reaction temperature is 60 °C.

The hydrogenation reaction rate increases with the increase of pressure in the initial stage, and then reaches a stable value within a certain pressure range. However, if the pressure is further increased, the reaction rate will drop significantly. This phenomenon may be due to the fact that the reactants to be hydrogenated are squeezed by hydrogen on the active sites of the catalyst surface as the pressure is further increased. Generally, suitable hydrogenation pressure is in the range from 1.0 to 3.0 MPa.

4.4.3 Uses of methylhexynediol and dimethylhexanediol

Methylhexynediol can be used as a surfactant and as an intermediate in the synthesis of herbicides. It is also used as a chelating agent for fiber and paper etchants, a metal phosphate processing reagent, a lubricant for electric wires and an intermediate for organic peroxides.

Dimethylhexanediol is an important raw material for high-efficiency and low-toxic pyrethroid pesticides. It is also used as a combination of new plastic cross-linking agents, surfactants, artificial musk and fragrances.

4.5 Methyl heptenone, dehydrolinalool and linalool

Methyl heptenone, also known as 6-methyl-5-hepten-2-one, 6-methyl-5-heptene-2-one, 2-methyl-2-hepten-6-one, sulcatone and prenylacetone, is a derivative of acetylene and acetone. CAS Registry Number: 110-93-0. Molecular formula: $C_8H_{14}O$, molecular weight: 126.1962. The chemical structure is

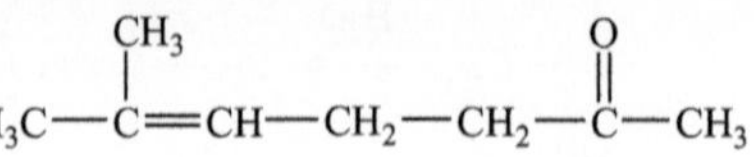

Dehydrolinalool, also known as 3,7-dimethyl-6-octen-1-yn-3-ol, is a derivative of methyl heptenone and acetylene. CAS Registry Number: 29171-20-8. Molecular formula: $C_{10}H_{16}O$, molecular weight: 152.2334. The chemical structure is

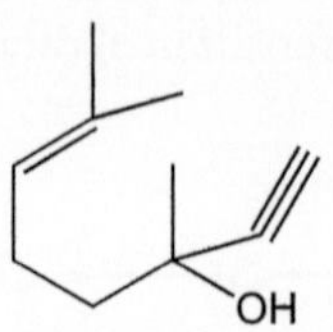

Linalool, also known as 3,7-dimethyl-1,6-octadien-3-ol, β-linalool, linalyl alcohol, 2,6-dimethyl-2,7-octadien-6-ol, 2,6-dimethyl-2,7-octadiene-6-ol, 2,6-dimethylocta-2,7-dien-6-ol, 3,7-dimethylocta-1,6-dien-3-ol, DL-3,7-dimethyl-3-hydroxy-1,6-octadiene, phantol, 3,7-dimethyl-1,6-octandien-3-ol, *R*/*S*-linalool, (*S*)-linalool, (+)-linalool, (−)-β-linalool, (*S*)-linalool, L-linalool, (±)-linalool and (+)-β-linalool, is a derivative of acetylene and acetone. CAS Registry Number: 78-70-6. Molecular formula: $C_{10}H_{18}O$, molecular weight: 154.2493. The chemical structure is

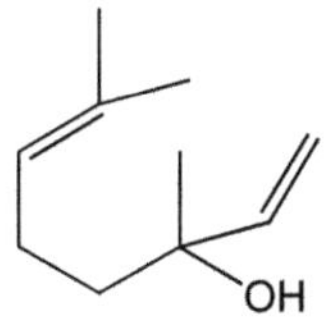

4.5.1 Properties of methyl heptenone

Methyl heptenone is a colorless or light yellow liquid with a lemon grass and isobutyl acetate-like aroma. It is miscible with alcohols and ethers but insoluble in water. The physical properties of methyl heptanone are summarized in Table 4.5. It is chemical active and from which a variety of products can be derived. Therefore, it is an important intermediate for the synthesis of medicines, flavors and fragrances.

Table 4.5: Physical properties of methyl heptenone.

Melting point/°C	−67.1	Flash point (open cup)/°C	50
Boiling point/°C	173.2	Density (20 °C)/$g \cdot cm^{-3}$	0.860
Vapor pressure/kPa		Enthalpy of vaporization/$kJ \cdot mol^{-1}$	
58.5 °C	1.33	106 °C	45.9
73.0 °C	2.40	117 °C	44.7
Heat capacity (25 °C)/$J \cdot mol^{-1} \cdot K^{-1}$	268.8	Refractive index (20 °C)	1.4390

4.5.2 Properties of dehydrolinalool and linalool

Dehydrolinalool is an acyclic monoterpene alcohol compound with a pleasant floral vanilla aroma and a spicy and lavender-like fragrance. Its boiling point is 89–93 °C at 1.6 kPa. The constant pressure heat capacity is 385.2 $J \cdot mol^{-1} \cdot K^{-1}$ at 313.55 K. The enthalpies of vaporization are 50.4 $kJ \cdot mol^{-1}$ at 407 K and 52.1 $kJ \cdot mol^{-1}$ at 421 K. It is used in small amounts for the production of chemical fragrances for daily use. Industrial-grade dehydrolinalool (content 92–96%) is mainly used to synthesize a series of perfumes such as linalool, nerolidol, citral and ionone, which are also important intermediates for the synthesis of vitamin A, E and K. Dehydrolinalool has a strong ability to mask bad odors, especially for disgusting hydrogen sulfide, garlic preparations and polysulfide odors. In addition, it also has the role of mosquito repellent. A sterol compound such as dehydrolinalool has excellent antibacterial activity against *Escherichia coli*, *Enterobacter faecalis*, *Proteus* and the like. Dehydrolinalool has high added value and is a fine chemical product with broad market prospects.

Linalool is a colorless liquid with a flash point of 76 °C. It is insoluble in water and glycerin, and can be miscible with ethanol and ether. Linalool is the main component of natural eucalyptus oil, galaxie oil, rosewood oil and so on, and all of them are levorotatory isomers, while those present in eucalyptus oil and sesame oil are dextrorotatory isomers. The physical properties of these two isomers are listed in Table 4.6.

Table 4.6: Physical properties of the two isomers.

Item	Dextrorotatory form	Levorotatory form
Boiling point/°C	198–200	198
Density (20 °C)/$g \cdot cm^{-3}$	0.8733	0.8622
Optical rotation (20 °C)	+19.3°	−20.1°

Most of the linalool and its esters have a beautiful and pleasant floral aroma. It has a chiral carbon atom with different aromas for linalool with different optical rotations. Its unique aroma quality makes it an important part of many fragrance formulas, which can be used to blend various flavors such as lily, clove and orange blossom. In addition, it is also an important raw material for the synthesis of linalool flavor compounds and vitamins E and A. Earlier, linalool was mainly obtained from many natural essential oils. With the increasing use and dosage of linalool and its derivatives, the production of essential oils cannot keep up with the development needs, so the production by synthetic methods has attracted attention. In 1919, Ruzika and Fornasir successfully used acetylene and acetone to synthesize linalool by condensation, ethynylation and hydrogenation. Since then, various synthetic methods have emerged, the most notable of which is the synthesis of linalool from a-pinene, which is rich in resources.

4.5.3 Synthesis of methylheptenone

There are three synthetic routes for methylheptenone according to the starting materials. One is from acetylene and acetone, the second is from isobutylene and the third is from isoprene. Each of these three routes has advantages and disadvantages. The acetylene–acetone process is relatively mature and the raw materials are abundant; therefore, it is the main method of industrial production. The disadvantages are many reaction steps required and high cost of raw material (methyl acetoacetate) used. Isobutylene and isoprene are important products in the C4 and C5 fractions of petroleum refining, and their sources are abundant. However, the process conditions of isobutylene process are harsh (high temperature and high pressure are required) and the yield of 6-methyl-5-hepten-2-one is low. Industrialization reports for the production of 6-methyl-5-hepten-2-one from isobutylene have not been reported in China. Although the reaction steps in isoprene route is less than that in acetylene–acetone route, the yield of 6-methyl-5-hepten-2-one for isoprene route is lower than that for acetylene–acetone route. The costs of the raw materials for isoprene route are low, which makes the isoprene route to have a certain competitiveness.

4.5.3.1 Manufacture of methylheptenone from acetylene and acetone

Acetone, the starting material, is subjected to ethynylation by acetylene in the presence of an alkaline catalyst to form 2-methyl-3-butyn-2-ol and successive partial hydrogenation in the presence of a Lindlar catalyst, followed by reaction with diketene or alkyl acetoacetate (ethyl acetoacetate or methyl acetoacetate) to form an ester derivative of acetoacetic acid: 2-methyl-3-buten-2-yl acetoacetate. Thereafter, the ester thus formed is further subjected to Carroll rearrangement to produce 6-methyl-5-hepten-2-one (Figure 4.4).

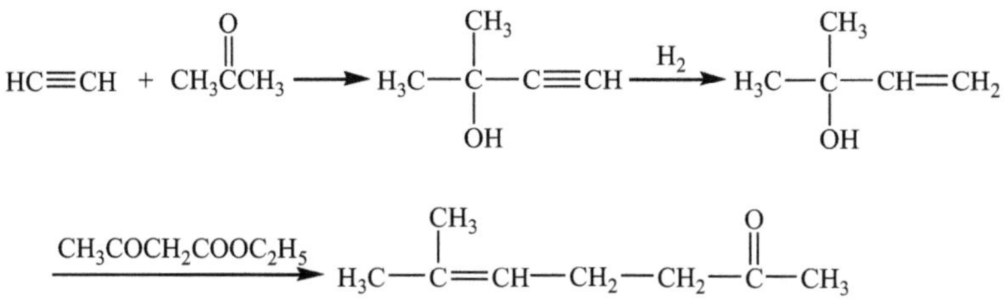

Figure 4.4: Reaction process for the manufacture of methyl heptenone from acetylene and acetone.

The route was first industrialized by Swiss company F. Hoffmann-La Roche. German company BASF has improved the process and applied it to produce 6-methyl-5-hepten-2-one. China's Southwest Chemical Industry Research Institute has also developed the process technology for the preparation of methylheptenone and linalool from acetylene and acetone.

The processes from acetylene and acetone to 2-methyl-3-butyn-2-ol and 2-methyl-3-buten-2-ol were described in Section 4.1.2. The downstream process is described here. 2-Methyl-3-buten-2-ol was mixed with an equal volume of toluene to form a solution. To the solution, sodium methoxide (1.2% based on mass of 2-methyl-3-buten-2-ol) was added and dissolved. Diketene with an excess of 10% over 2-methyl-3-buten-2-ol was continuously added to the solution within 2 h. The reaction temperature was maintained at 25–30 °C by external cooling. After the addition of diketene, the solution was continuously stirred for 5 h. Then the mixture was treated by washing successively with dilute sulfuric acid, saturated sodium bicarbonate solution and finally with water until neutral. 2-Methyl-3-buten-2-yl acetoacetate with a yield of 97% was collected by distillation at reduced pressure [8].

Aluminum isopropoxide (1.7% based on mass of 2-methyl-3-buten-2-yl acetoacetate) was added to 2-methyl-3-buten-2-yl acetoacetate in a stirring kettle equipped with mechanical stirrer and condenser connected at the top. The mixture was stirred vigorously and was heated to 140–160 °C for 5 h. 6-Methyl-5-hepten-2-one with a yield of 83% was collected by distillation at reduced pressure. The total yield of 6-methyl-5-hepten-2-one was about 80.5% based on 2-methyl-3-buten-2-ol.

Since diketene is unstable, a high safety requirement for the operation must be ensured. For safety reasons, the process was improved by using ethyl acetoacetate or methyl acetoacetate instead of diketene. The reaction was carried out in a kettle connected to a distillation column. Aluminum triphenolate was used as catalyst and dissolved in the mixture of 2-methyl-3-buten-2-ol and methyl acetoacetate. The solution was heated to 145 °C, and then 2-methyl-3-buten-2-ol and methyl acetoacetate with a molar ratio of 1.25:1 were continuously and uniformly added to the solution in the reaction kettle during a period of 9.5 h. At the same time, the methanol generated was continuously distilled off. The yield of 6-methyl-5-hepten-2-one was 89.5% based on methyl acetoacetate [9].

4.5.3.2 Manufacture of 6-methyl-5-hepten-2-one from isobutylene

Germany company BASF used isobutylene, acetone and formaldehyde as raw materials to synthesize α-methylheptenone at 250 °C and 30 MPa. Subsequently, α-methylheptenone was heated and converted to 6-methyl-5-hepten-2-one in the presence of palladium and carbonyl iron catalyst. Sun et al. [10] synthesized methylheptenone by one-step reaction from isobutylene, acetone and formaldehyde. Under the reaction conditions of reaction pressure of 30 MPa, reaction temperature of 310–320 °C, isobutylene:acetone:formaldehyde 5:4:1 (molar ratio) and reaction residence time of 115 h, the yield of methylheptenone was 34% based on formaldehyde and the space–time yield is 35 g/(h · L). The results show that the addition of phosphoric acid or caustic soda cannot play a catalytic role; the lower the reaction temperature, the higher the ratio of α-methylheptenone to β-methylheptenone. Since the process is a condensation reaction between three molecules, many side reactions occur, leading to high requirements for product separation and purification.

The reaction process is shown in Figure 4.5.

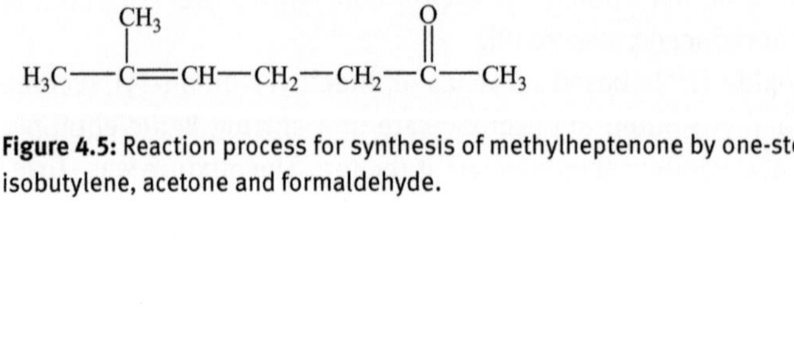

Figure 4.5: Reaction process for synthesis of methylheptenone by one-step reaction from isobutylene, acetone and formaldehyde.

4.5.3.3 Manufacture of 6-methyl-5-hepten-2-one from isoprene

This process was first proposed by the French company Rhodia. Isoprene reacts with hydrogen chloride to form isopentenyl chloride with a yield of 67% based on isoprene. The latter reacts with acetone to form 6-methyl-5-hepten-2-one. The first step is a batch reaction process in a waterless system. The Japanese company Kuraray had improved the second step by using a quaternary ammonium salt as a phase transfer catalyst for continuous reaction in an aqueous sodium hydroxide solution. The company built a production unit for 6-methyl-5-hepten-2-one based on the improved technology. The reaction process is as follows:

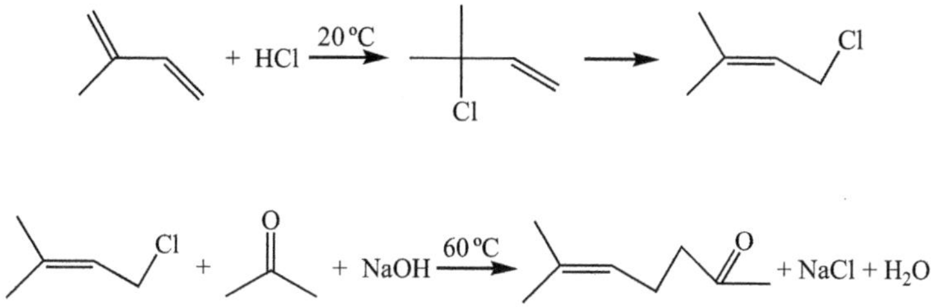

The phase transfer catalysts such as cetyltrimethylammonium bromide and benzyltriethylammonium chloride can be used for the condensation reaction. Benzyltriethylammonium chloride is more suitable because its solubility is higher and the solid waste generated is less than that of cetyltrimethylammonium bromide. The reaction was carried out at a catalyst dosage of 0.4% based on isopentenyl chloride, isopentenyl chloride:acetone:NaOH solution (48–51%) = 1:3.9:6.5 (mass ratio) and the reaction temperature of 60–61 °C for 3 h, the yield of 6-methyl-5-hepten-2-one was 65% based on isopentenyl chloride.

Since the source of isoprene is abundant, it is a potential process for 6-methyl-5-hepten-2-one. The Chemical Engineering Research Institute of SINOPEC, Shanghai, has also developed the process technology for the manufacture of 6-methyl-5-hepten-2-one from isoprene and completed a pilot test in 1997.

4.5.4 Process for manufacture of dehydrolinalool and linalool

Dehydrolinalool is obtained by ethynylation of methylheptenone with acetylene in the presence of potassium hydroxide and liquid ammonia:

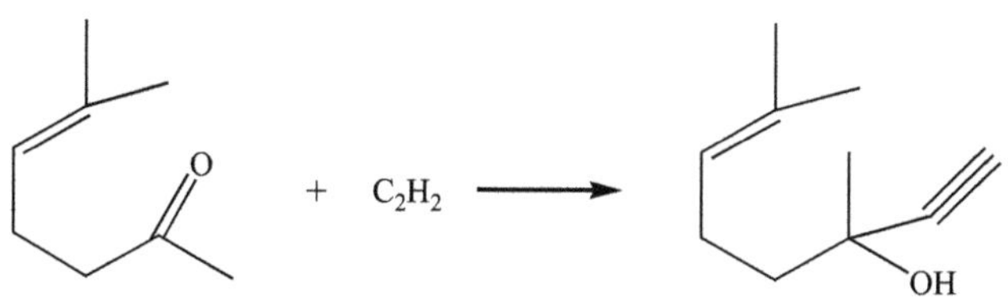

The specific process is as follows: open the valves of ammonia and acetylene gas cylinders, respectively, add ammonia gas and acetylene to the mixed gas cabinet, and adjust the ratio of the two gases so that the content of acetylene in the mixed gas cabinet is about 20–25%. The gas mixture is compressed to 2.0–2.5 MPa by a compressor, liquefied in a high-position cooling tower and enters the reaction tower through a bottom discharge pipe, and a methylheptene solution containing potassium hydroxide is introduced into the reaction tower through a metering pump. The reaction temperature is 35–40 °C, the reaction pressure is 2.0–2.5 MPa and the reaction time is 2–3 h. The reaction liquid enters the flash tank through the discharge tube of the reaction tower, and the flashing temperature is 50–70 °C. Excess acetylene and ammonia are flashed and recycled back to the mixing gas cabinet, and the reaction product overflows from the bottom of the flash tank to the storage tank. By distillation under reduced pressure, dehydrolinalool having a mass fraction of more than 98% can be obtained and the yield can reach 92–94% [11].

Dehydrolinalool can also be produced by reaction of methylheptenone with acetylene in *N*-methylpyrrolidone in the presence of sodium hydroxide catalyst under mild conditions. However, sodium hydroxide is used in large quantities (molar ratio of sodium hydroxide to methylheptenone is as high as 3.75) [12].

Synthesis of linalool by selective catalytic hydrogenation of dehydrolinalool is shown as follows:

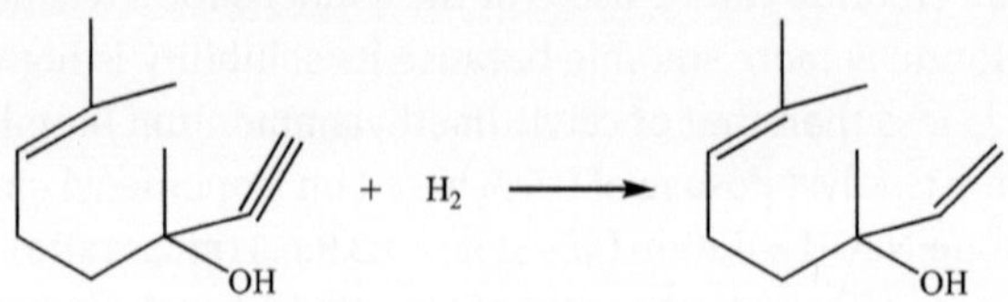

Generally, elements such as palladium, rhodium, platinum and nickel are selected as catalysts, and the carrier is usually Al_2O_3 or $CaCO_3$. It is further modified with lead, bismuth, manganese and pyridine, quinoline and sulfur compounds. Both the conversion of dehydrolinalool and the yield of linalool were more than 99% at a reaction temperature of 60 °C and a pressure of 0.4 MPa.

Research on the kinetics of liquid-phase catalytic hydrogenation of dehydrolinalool in a batch autoclave using Raney–Ni catalyst indicated first-order reaction for both dehydrolinalool and hydrogen. Yet a side reaction of further hydrogenation of linalool to a by-product dihydrolinalool showed a zero-order reaction for the linalool and a first-order reaction for hydrogen [13].

4.5.5 Uses of methylheptenone, dehydrolinalool and linalool

Methylheptenone is an intermediate for the preparation of dehydrolinalool, linalool, citral and pseudoionone, which can be further used for synthesis of vitamin A, vitamin E, vitamin K1 and various flavors and fragrances.

Dehydrolinalol is an important intermediate for the preparation of β-ionone and vitamin E. It can be used to produce a series of fragrances such as linalool and nerolidol. Furthermore, dehydrolinalool is a fragrance ingredient used in decorative cosmetics, fine fragrances, shampoos, toilet soaps and other toiletries as well as in non-cosmetic products such as household cleaners and detergents [14].

Linalool is a fragrance and flavor substance [15]. It is a blending raw material for various artificial essential oils. Its unique aroma quality makes it an important part of many fragrance formulas, which can be used to blend various flavors such as lily, clove and orange blossom. In addition, it is also an important raw material for the synthesis of geranyl acetone and β-ionone, isophytol, vitamin A and E and β-carotene.

4.6 Geranyl acetone, dehydronerolidol and nerolidol

Geranyl acetone, also known as 6,10-dimethyl-5,9-undecadien-2-one; 6,10-dimethyl-5,9-undecadiene-2-one, dihydropseudoionone, α, β-dihydropseudoionone, is a derivative of linalool. CAS Registry Number: 689-67-8. Molecular formula: $C_{13}H_{22}O$, molecular weight: 194.3132. The chemical structure is

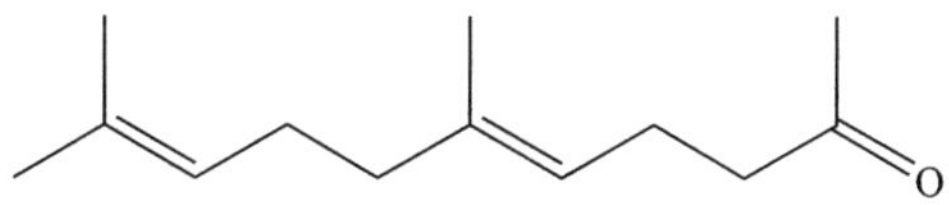

Dehydronerolidol, also known as 3,7,11-trimethyldodeca-6,10-dien-1-yn-3-ol, is a derivative of geranyl acetone. CAS Registry Number: 2387-68-0. Molecular formula: $C_{15}H_{24}O$, molecular weight: 220.3505. The chemical structure is

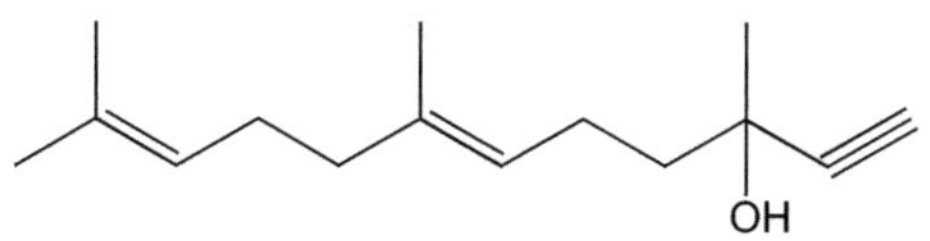

Nerolidol, also known as peruviol, 3,7,11-trimethyl-1,6,10-dodecatrien-3-ol, 3,7,11-trimethyl-1,6,10-dodecatriene-3-ol, 3,7,11-trimethyldodeca-1,6,10-trien-3-ol, 3-hydroxy-3,7,11-trimethyl-1,6,10-dodecatriene, is a derivative of dehydronerolidol. CAS Registry Number: 7212-44-4. Molecular formula: $C_{15}H_{26}O$, molecular weight: 222.3663. The chemical structure is

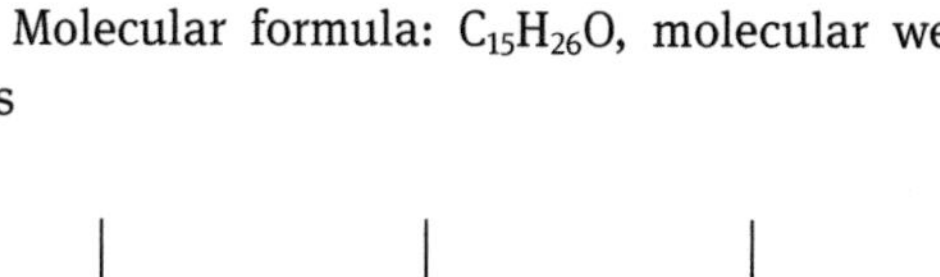

4.6.1 Properties of geranyl acetone, dehydronerolidol and nerolidol

Geranyl acetone is a light yellow to colorless, transparent and oily liquid with an aroma of magnolia. Its boiling point is 250 °C (101.3 kPa), and the flash point is 101 °C. The density is 0.8729 g cm^{-3} and the refractive index is 1.4674 at 20 °C. Geranyl acetone has a (*Z*)-isomer, neryl acetone or (*Z*)-geranyl acetone.

Dehydronerolidol is a colorless and transparent liquid having a boiling point of 318.5 °C (101.3 kPa), a flash point of 289.8 °C, a refractive index of 1.49 (25 °C) and a density of 0.902 g cm^{-3}.

Nerolidol is a colorless to straw yellow syrupy oily liquid with a faint scent similar to rose and apple, accompanied by a very sweet, fresh, long-lasting aroma. There are (*Z*)- and (*E*)-isomers of nerolidol, which differ in the geometry on the central double bond. The physical properties of nerolidol (mixture of *Z*/*E* isomers) are listed in Table 4.7.

Table 4.7: Physical properties of nerolidol (mixture of *Z*/*E* isomers).

Melting point/°C	−75	Vapor pressure/kPa	
Boiling point/°C	276	102 °C	0.0532
Density (25 °C)/g · cm^{-3}	0.8750	114 °C	0.1
Refractive index (20 °C)	1.4790	122 °C	0.4
Flash point (open cup)/°C	110	145–146 °C	1.6

4.6.2 Production process of geranyl acetone, dehydronerolidol and nerolidol

4.6.2.1 Process for manufacture of geranyl acetone

Although the natural geranyl acetone is found in tomatoes, mint and passion fruit, the product is industrially produced by chemical synthesis. The chemically synthesized product is a mixture of (*Z*)- and (*E*)-isomers. The (*Z*)-isomer is called neryl acetone and the (*E*)-isomer is called geranyl acetone. There are three routes for manufacture of geranyl acetone: the first one is that myrcene reacts with methyl acetoacetate in the presence of a precious metal catalyst and subsequently the intermediate is subjected to hydrolysis and decarboxylation to obtain geranyl acetone. However, the catalyst used in the reaction process is very expensive, which is not conducive to industrialization. The second synthesis route is that myrcene and hydrogen chloride undergo an addition reaction to obtain geranyl chloride and its isomers, and then the as-prepared chlorides react with ethyl acetoacetate followed by hydrolysis and decarboxylation to generate geranyl acetone. Although the use of precious metal catalysts can be avoided, a large amount of wastewater is generated in the second synthesis route. The third synthesis route is that geranyl acetone is directly prepared by the Carroll reaction using linalool and ethyl acetoacetate as raw materials. Currently, the third route is widely used, and the commonly used catalysts are aluminum alkyl, aluminum isopropoxide, sodium ethoxide, sodium dihydrogen phosphate, disodium hydrogen phosphate and the like. The reaction equation is as follows:

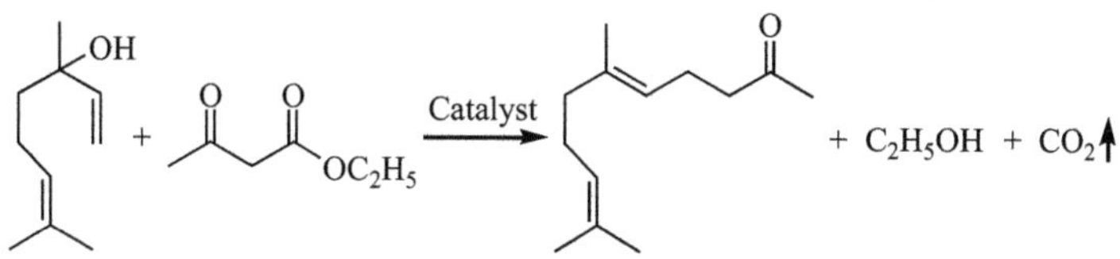

For example, using sodium dihydrogen phosphate as the catalyst, the reaction was carried out at 170 °C for 8 h, with a catalyst dosage of 3% based on the mass of linalool and a molar ratio of linalool to ethyl acetoacetate of 1:2, resulting in a total yield of geranylacetone and nerylacetone up to 97.5% [16]. The ratio of geranylacetone to nerylacetone is approximately 6:4.

4.6.2.2 Process for production of dehydronerolidol

In the presence of an alkali metal hydroxide catalyst, geranyl acetone reacted with acetylene under atmospheric pressure to form dehydronerolidol with a yield of 85.7%. The reaction equation is as follows:

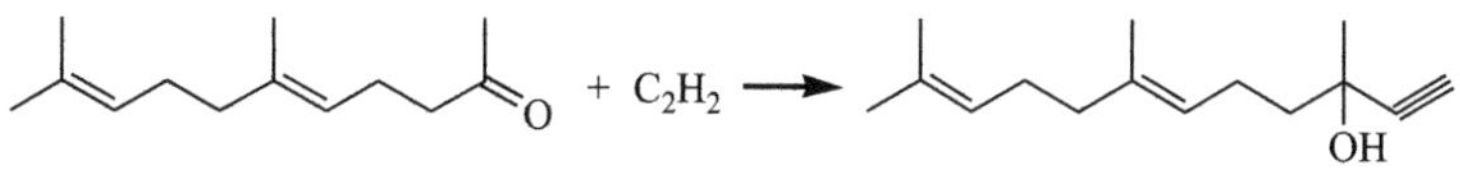

The alkali metal hydroxide, usually a fine and uniform powdered potassium hydroxide, is dispersed in the organic phase. Potassium hydroxide is in excess of the raw material geranyl acetone. The solvent type has a great influence on the reaction, and a solvent such as toluene or tetrahydrofuran can be used. Among them, tetrahydrofuran has the best solvent effect because acetylene has a large solubility in tetrahydrofuran, and solid potassium hydroxide can be dissolved and dispersed well in it. If no solvent is introduced, the reaction system will be thick and uneven, resulting in the production of significant amounts of by-products and thus a low yield of the target product dehydronerolidol. The water in the solvent is unfavorable to the ethynylation reaction, so it should be removed before use.

In general, the reaction is carried out for 4–8 h at a temperature between 0 and −5 °C. At low temperature, the solubility of acetylene in the reaction system is relatively large, which is favorable to the enhancement of geranyl acetone conversion. Moreover, the selectivity to dehydronerolidol is higher and the by-products are fewer at low temperature. As the reaction temperature is raised to room temperature or higher, the by-products are significantly increased.

4.6.2.3 Process for manufacture of nerolidol

Nerolidol was prepared by selective partial hydrogenation of dehydronerolidol in the presence of a catalyst:

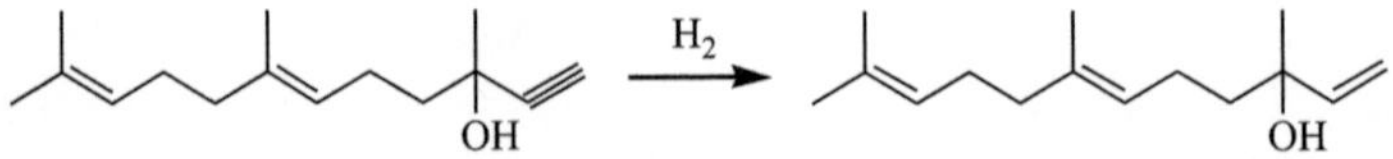

An autoclave having an internal volume of 3 L was charged with 630 g of dehydronerolidol, 270 g of hexane and 0.22 g of a Lindlar catalyst (0.25% Pd supported on calcium carbonate). Hydrogenation was carried out for 4 h under hydrogen pressure of 0.5–0.8 MPa (gauge pressure) at a temperature of 25–43 °C. Thereafter, the catalyst was filtered off, and the resultant filtrate was concentrated with a rotary evaporator to give 630 g of crude nerolidol. The conversion of dehydronerolidol was 95.5%, and the selectivity to nerolidol was 94.6%.

Hydrogenation of dehydronerolidol was also carried out at atmospheric or nearly atmospheric pressure in the presence of a Lindlar catalyst with 5% Pd and quinoline. Toluene was applied as the solvent instead of hexane. Under optimized conditions (catalyst dosage: 0.075% based on the mass of dehydronerolidol; ratio of solvent to dehydronerolidol: 2:1; reaction temperature: 60 °C; reaction time: 5 h), the yield of nerolidol reached to 96. 2%.

4.6.3 Uses of geranyl acetone, dehydronerolidol and nerolidol

Geranyl acetone is a fragrance and flavor substance, which is widely used in foods, beverages, cosmetics, household cleaners and detergents. It can be used to formulate a variety of flavors such as apple, banana and citrus essence. When used in the preparation of citrus essence, it can effectively prevent the bitter and astringent taste of real citrus. Geranyl acetone can be used in both rose-flavored flavors as an auxiliary flavor in non-flora. The combination of geranyl acetone and lavender and fruity is very effective, which can play the role of refreshing and natural aroma and make the floral fragrance more mellow. In addition, geranyl acetone is also used in the cigarette industry to enhance the coordination of smoke aroma.

Geranyl acetone is a key intermediate for the synthesis of dehydronerolidol and farnesol and is also a key intermediate in the production of isophytol. Isophytol is an important side chain of synthetic vitamin E, and an increase in the demand for vitamin E is bound to expand the production of geranyl acetone.

Dehydronerolidol is mainly applied to production of nerolidol by partial reduction of the acetylenic bond in its molecule.

Nerolidol is a fragrance ingredient used in a variety of flavor formulations [17]. It is used in food and cosmetics. Furthermore, nerolidol is applied for the manufacture of farnesyl acetone and finally production of vitamin E and K_1. Nerolidol is also a key intermediate for the synthesis of squalene, which is a deep-sea fish oil health product with broad market prospects.

4.7 Isophytol

Isophytol, also known as 3,7,11,15-tetramethyl-1-hexadecen-3-ol, is a derivative of acetylene and acetone. CAS Registry Number: 505-32-8. Molecular formula: $C_{20}H_{40}O$, molecular weight: 296.5310. The chemical structure is

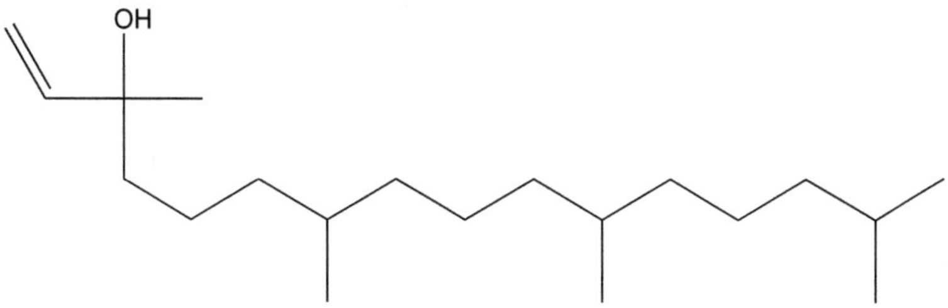

4.7.1 Properties of isophytol

Isophytol is a colorless viscous liquid with a delicate dry leaf, tealike and slightly carmellic sweet odor. It is insoluble in water but soluble in organic solvents such as alcohols and ketones. It has the dual nature of olefins and alcohols. The physical properties of isophytol are listed in Table 4.8.

Table 4.8: Physical properties of isophytol.

Boiling point/°C	334.88	Density (25 °C)/g · cm^{-3}	0.8410
Vapor pressure/kPa		Flash point (open cup)/°C	143.5
107–110 °C	1.4	Refractive index (20 °C)	1.4571
125–128 °C	8.2	Heat capacity (20 °C)/J · mol^{-1} · K^{-1}	729.5
Density (20 °C)/g · cm^{-3}	0.8458	Enthalpy of vaporization (180 °C)/kJ · mol^{-1}	67.0

4.7.2 Production process of isophytol

According to the different key intermediates in the preparation process, the industrial production processes of isophytol include pseudoionone route and linalool route. The pseudoionone route using *Litsea cubeba* oil as a raw material is a traditional method for preparation of isophytol. However, due to the limitation of natural resource of *Litsea cubeba* oil, the production scale is small and cannot meet the market demand. Currently, linalool route is the most widely used process for isophytol production in the world. Various starting materials have been used for linalool synthesis, including acetylene–acetone (Roche process), isobutylene–formaldehyde–acetone (BASF process), isoprene and natural resource turpentine.

Roche process for production of isophytol was developed by Hoffmann-La Roche Company in Switzerland, which enabled the company to take the lead in mastering the large-scale production technology of vitamin E in the world and occupy a leading position in the market for a long time. In the process, acetylene and acetone were used as starting materials and underwent more than ten steps of reactions via methyl butynol, methyl heptenone, dehydrolinalool, linalool, dehydronerolidol, nerolidol, farnesylacetone, phyton and dehydroisophytol to produce isophytol. Various intermediates and isophytol produced by the Roche process are of good qualities. The production operations are easy to control and no corrosion to production equipment occurs. Therefore, the Roche process has been adopted by the vast majority of isophytol manufacturers in the world for production of isophytol.

Figure 4.6 shows the reaction route from nerolidol to isophytol. Since the preparation of nerolidol has been described in Section 4.6, this section only focuses on the downstream reaction processes starting from nerolidol. Nerolidol undergoes a Carroll reaction with methyl acetoacetate in the presence of aluminum isopropoxide catalyst at a temperature of 170 °C for 3 h. The molar ratio of nerolidol and methyl acetoacetate is 1:1.3, and the catalyst dosage is 4% based on the weight of nerolidol. After the completion of the reaction, the reaction solution is distilled under reduced pressure, and the fraction at a temperature of 125–126 °C and a pressure of 1.33 mbar is collected. The yield of farnesyl acetone was 95.8% with a purity of 99.4% [18].

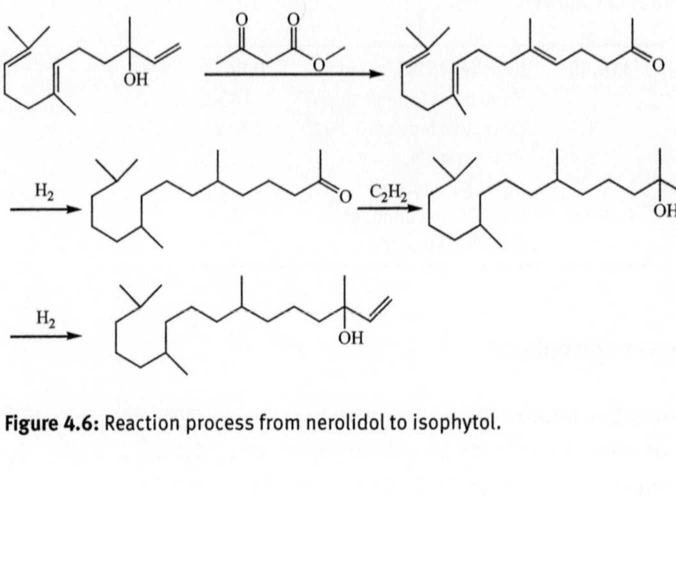

Figure 4.6: Reaction process from nerolidol to isophytol.

Farnesylacetone is dissolved in methanol and hydrogenated in the presence of Pd/C catalyst at room temperature under 0.141 MPa for 24 h. After removal of the catalyst by filtration, the reaction solution is concentrated under reduced pressure to remove the solvent, and the residue is distilled under reduced pressure to collect a fraction at 152–153 °C and 2.6 mbar. The yield of phyton is 95.4% with a purity of 99.2%.

Phyton underwent an ethynylation reaction with acetylene in potassium hydroxide/liquid ammonia system at 4–6 °C for 2.5 h. An aqueous 25% ammonium sulfate was added to stop the reaction. The ammonia and unreacted acetylene were recovered. The dehydroisophytol was isolated by extraction with hexane and distillation. The conversion of phyton was 95.4%. The resultant dehydroisophytol was hydrogenated in the presence of a Lindlar catalyst under hydrogen pressure of 0.4 MPa and at a temperature between 30 and 60 °C for 8 h. After filtration of the catalyst and evaporation of the solvent, isophytol was obtained by vacuum distillation under 2–3 mmHg and collection of a fraction at 148–160 °C. The yield of isophytol was 75.7% based on phyton [19].

4.7.2.1 Preparation of isophytol from *Litsea cubeba* oil

Litsea cubeba oil is a kind of natural essential oil extracted from fresh fruit of *Litsea cubeba*, which is yellowish and contains more than 60% of citral. China is rich in *Litsea cubeba* oil resources which can be used to prepare isophytol. Figure 4.7 shows the reaction process

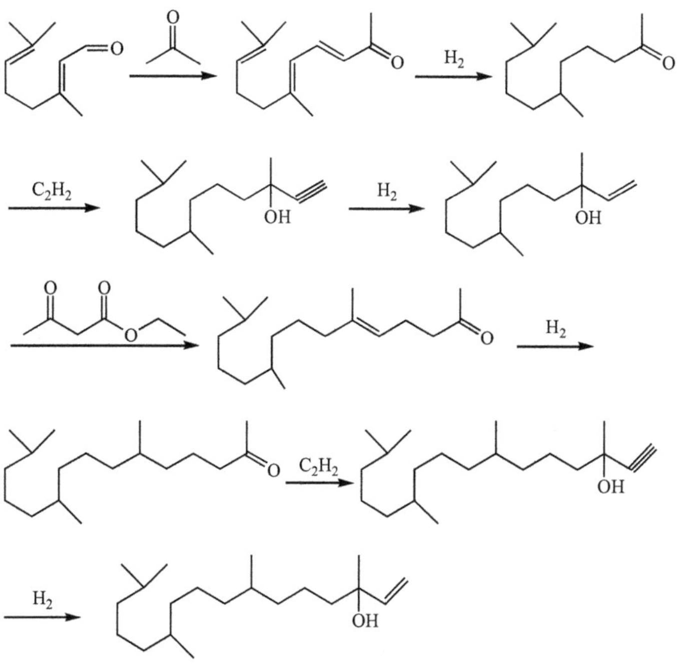

Figure 4.7: Litsea cubeba oil as the starting material via pseudoionone to isophytol.

using *Litsea cubeba* oil as the starting material via pseudoionone to isophytol. The citral in *Litsea cubeba* oil underwent a condensation reaction with acetone to yield pseudoionone. The resulting pseudoionone was hydrogenated and then reacted with acetylene to generate a C15 alkynol. Semihydrogenation of the C15 alkynol was followed by a Carroll reaction with ethyl acetoacetate leading to phyton. Ethynylation of the phyton with acetylene gave dehydroisophytol, which was partially hydrogenated to obtain isophytol.

4.7.3 Uses of isophytol

Isophytol is a terpenoid alcohol that is used as a fragrance[20] and as an intermediate in the production of vitamin E and K1. Among them, the amount of isophytol used in production of vitamin E is the largest, followed by that used in the production of vitamin K1, while only about 0.1% of isophytol is used for fragrance and flavoring. Vitamin E is widely used in medicine, food, animal feed, beauty cosmetics and other aspects. As an antioxidant, vitamin E can effectively inhibit the oxidation reaction in the body, especially to protect the cell membrane from the attack of free radicals produced by unsaturated ester compounds due to peroxidation. When the body is deficient in vitamin E, it will directly affect some of the normal functions of the body, especially in the reproductive, bone and nerve systems. Moreover, the addition of vitamin E to foods can play a role in preservation. Vitamin E can decisively protect the skin from the erosion of free radicals. Therefore, vitamin E is added to many cosmetic products. The effect is to promote skin metabolism while preventing pigmentation, to effectively improve skin elasticity and to play a role in beauty and antiaging. Vitamin E is also used as a feed additive, showing multiple effects: antioxidation and catalyzing the growth of livestock and poultry. It is of great significance in animal immunization, disease prevention, promotion of livestock and poultry reproduction and egg production.

4.8 3-Methyl-2-penten-4-yn-1-ol

3-Methyl-2-penten-4-yn-1-ol, also known as C_6 alcohol or 3-methylpent-2-en-4-yn-1-ol, is a derivative of methyl vinyl ketone and acetylene. CAS Registry Number: 105-29-3. Molecular formula: C_6H_8O, molecular weight: 96.1271. There are two isomers: *cis*- and *trans*-C_6 alcohol. The chemical structures are as follows:

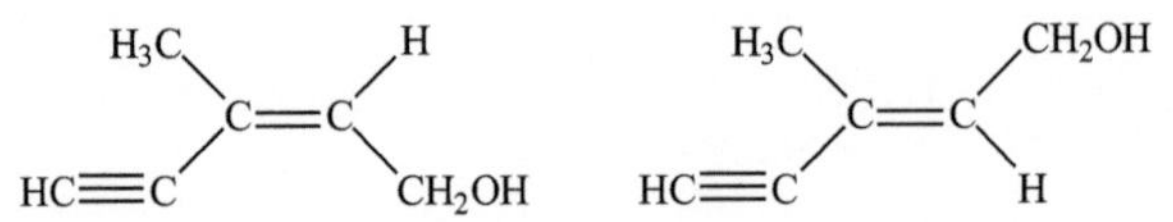

cis (Z)- 3-methyl-2-penten-4-yn-1-ol

trans (E)-3-methyl-2-penten-4-yn-1-ol

4.8.1 Properties of 3-methyl-2-penten-4-yn-1-ol

The cis isomer is an oily liquid with a boiling point of 65 °C (1.253 kPa) and a refractive index of 1.4820 (20 °C). The trans isomer is also an oily liquid with a boiling point of 73 °C (1.253 kPa) and a refractive index of 1.4934 (20 °C). Both isomers are heat-sensitive substances and easy to be polymerized because of double bonds and triple bonds in their molecular structures.

4.8.2 Production process of 3-methyl-2-penten-4-yn-1-ol

In the presence of metal calcium, methyl vinyl ketone reacts with acetylene to produce *iso*-C_6 alcohol (3-methyl-1-penten-4-yn-3-ol) which can be converted to C_6 alcohol via a rearrangement reaction catalyzed by sulfuric acid:

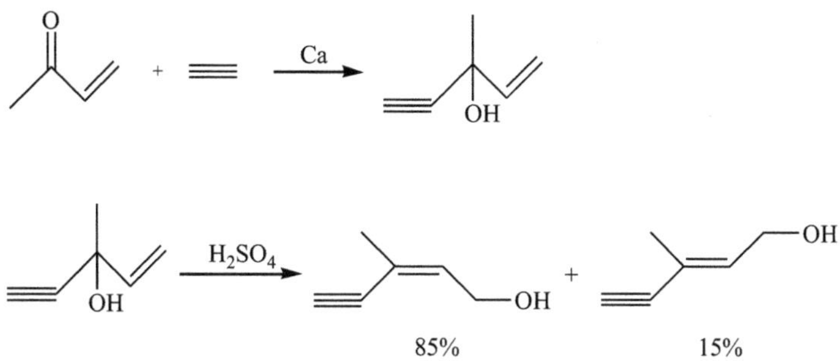

First, liquid ammonia is added to the reaction vessel, at a temperature below −40 °C. Metal calcium is added and simultaneously acetylene is introduced into the liquid ammonia which is under stirring. Calcium acetylide is formed and dispersed in the liquid ammonia solvent. The reaction mixture is cooled to −60 °C, then methyl vinyl ketone is added while acetylene is still bubbled therethrough at a moderate rate. The temperature of the reaction mixture is controlled at −55 °C for 25 minutes. The temperature was raised to −5 °C within 90 minutes in order to evaporate the ammonia. Equal amounts of diethyl ether and water are added followed by the addition of a 50% aqueous solution of acetic acid until neutral. During the additions, the temperature is controlled below 15 °C. The mixture is allowed to stand for separating into two layers : a water layer and an organic layer. The water layer is extracted twice with diethyl ether. The combined organic layers are washed with aqueous solution of sodium bicarbonate, dried over anhydrous magnesium sulphate and then distilled after recovery of diethyl ether. The yield of iso-C_6 alcohol is 71% based on methyl vinyl ketone.

115 g of iso-C6 alcohol is dissolved in 250 ml solvent (chloroform or isopropyl ether) under stirring, and 300 mL of dilute sulfuric acid (21.5%) is added to the solution. The reaction is carried out at a temperature of 55 °C for 2 h. After the reaction,

the reaction mixture is cooled and allowed to stand for separating into two layers. The organic layer is washed twice with aqueous solution of sodium bicarbonate (100 mL×2), dried over anhydrous magnesium sulphate and then distilled after recovery of the solvent. The yield of C_6 alcohol is 86.6% based on iso-C_6 alcohol and the content of cis isomer in the C_6 alcohol is 85.1%.

Sulfuric acid-catalyzed allyl rearrangement of 3-methyl-1-penten-4-yn-3-ol produces an isomeric mixture of (*Z*)- and (*E*)-3-methyl-2-penten-4-yn-1-ol. The cis isomer can be cyclized under certain conditions to form 2,3-dimethylfuran with a large exotherm. Both the isomers are heat-sensitive compounds and tend to polymerize when heated.

Kinetics study indicated that the reaction rate constants for the allyl rearrangement depend on acid concentration, the presence of solvent and temperature [21, 22]. The translocation rate is proportional to the concentration of sulfuric acid and increases by 2 to 4 times when using water instead of benzene as the solvent. However, the ratio of cis and trans isomers is enhanced when water is replaced by an organic solvent. For example, when the reaction is carried out with a sulfuric acid concentration of 20.5% and volume ratio of 1:2:3 (3-methyl-1-penten-4-yn-3-ol:dilute sulfuric acid:solvent) at 55 °C for 3 h, a ratio of about 85:15 (cis to trans) can be reached with isopropyl ether as the solvent, whereas the ratio is about 78:22 with water as the solvent.

Although sulfuric acid has high activity as a catalyst, it will generate a large amount of waste acid which cannot be recycled, resulting in environmental pollution. Furthermore, the use of sulfuric acid as a catalyst leads to severe equipment corrosion. In order to overcome the above shortcomings, acidic cation exchange resin was applied to replace sulfuric acid for the allyl rearrangement of 3-methyl-1-pentene-4-yn-3-ol. The reaction products are easy to separate from the catalyst, and the process can be carried out continuously. However, it needs a higher reaction temperature because the catalytic activity of the resin is weaker than that of sulfuric acid, leading to polymerization occurring on the inner surface of the resin micropores and thus a significant decrease in catalytic activity and selectivity after a period of operation.

Recently, acidic ionic liquids are used as catalysts to achieve allylic rearrangement of 3-methyl-1-pentene-4-yn-3-ol [23]. For example, $[HNMP]HSO_4$, an acidic ionic liquid formed by the reaction of NMP and sulfuric acid in a ratio of 1:1, is very effective with an increased proportion of cis isomer to 88% after reaction at 55 °C for 6 h. The ionic liquid is not volatile and can be easily separated from the reaction products by distillation. The ionic liquid can be recycled and there is no significant reduction in catalytic activity and selectivity after four cycles. However, the performance of the catalyst is significantly reduced when it is reused more than 4 times. This is probably because the ionic liquid is contaminated. The performance of the ionic liquid can be recovered by washing with petroleum ether or dichloromethane to remove the contaminants.

In industry, the reaction products of the allyl rearrangement usually contain cis isomer (78–85%), trans isomer (10–15%), light fractions (3–5%) and other impurities (2–4%). The trans isomer is always present in the reaction mixture. Optimization of reaction conditions such as temperature, solvent type and acid concentration cannot completely eliminate the formation of trans isomer. The cis isomer is an important intermediate for the manufacture of vitamin A. The trans isomer cannot be used to produce vitamin A. The cis and trans isomers can be separated from each other by fractional distillation [24].

Since the boiling points of cis and trans C_6 alcohols are not much different, so a large number of plates are required to separate each other by fractional distillation. An increase in the number of plates will increase the pressure drop of the tower, leading to the increase in the boiling point of the C_6 alcohols. Therefore, a higher operating temperature is required in the fractional distillation. However, C_6 alcohols are heat-sensitive compounds and easy to polymerize or decompose because of acetylenic and ethylenic bonds in their molecular structures.

The use of a falling film reboiler and steam distillation reduces the operating temperature in the fractional distillation, reduces the decomposition and polymerization of C_6 alcohols and improves the safety production factor.

Figure 4.8 displays the process flow diagram for continuous steam distillation of 3-methyl-2-penten-4-yn-1-ol. The chloroform solution of crude C_6 alcohols was preheated by feed preheater (2) and then introduced into the solvent removal tower (3) for fractional distillation. The solvent separated at the top of the tower is condensed by a condenser (4) and collected in the solvent receiving tank (7). The bottom product of the solvent removal tower (3) is collected in the intermediate tank (9) and preheated by another preheater (10) before being fed to the middle section of the product distillation tower (11). An appropriate amount of steam is introduced into the top of the falling film reboiler (16), and steam distillation was performed under negative pressure. Wire web packings with high efficiency and low pressure drop are applied in the distillation tower. The vapor distilled from the top of the product distillation tower (11)

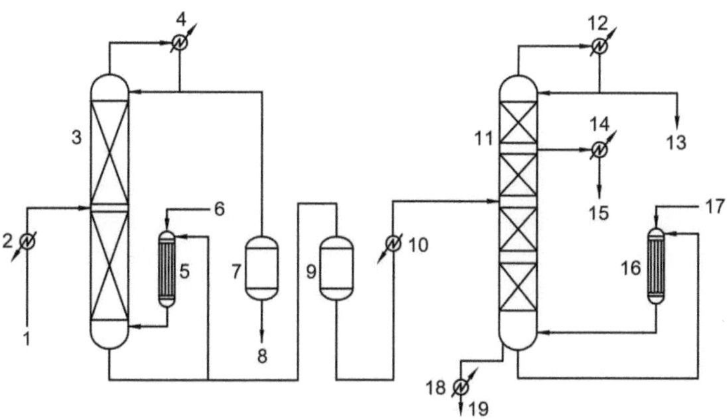

1. Crude product of C_6 alcohol containing solvent; 2, 10. Feed preheater; 3. Solvent removal tower; 4. Solvent condenser; 5, 16. Falling film reboiler; 6, 17. Steam; 7. Solvent receiving tank; 8. Solvent and water; 9. Intermediate tank; 11. Product distillation tower; 12. Light fraction condenser; 13. Light fractions and water; 14. Cooler; 15. cis C_6 alcohol; 16. Reboiler; 17. Steam; 18. Heavy fraction cooler; 19. trans C_6 alcohol.

Figure 4.8: Process flow diagram for continuous steam distillation of 3-methyl-2-penten-4-yn-1-ol.

is condensed by condenser (12) to obtain a mixture of light fractions and water, and some of them are withdrawn as the overhead product by controlling a certain reflux ratio. Final product of *cis*-C_6 alcohol (15) is withdrawn as a liquid side stream in the rectifying section of the tower. The bottom product (19) is withdrawn and cooled by a cooler (18) to obtain trans isomer which contains a small amount of oligomers. The yield of *cis*-C_6 alcohol is over 96% and the purity is more than 99%.

4.8.3 Uses of 3-methyl-2-penten-4-yn-1-ol

(*Z*)-3-Methyl-2-penten-4-yn-1-ol is a pharmaceutical intermediate for the production of vitamin A. Its stereoisomer, (*Z*)-3-methyl-2-penten-4-yn-1-ol, is also a useful intermediate for the manufacture of astaxanthin, zeaxanthin and further carotenoids [25].

4.9 Vitamin A

Vitamin A, also known as retinol, all-*trans*-retinol, all-*trans*-retinyl alcohol, *trans*-retinol and (all-*E*)-3,7-dimethyl-9-(2,6,6-trimethyl-1-cyclohexen-1-yl)-2,4,6,8-nonatetraen-1-ol, is a derivative of β-ionone and acetylene. CAS Registry Number: 68-26-8. Molecular formula: $C_{20}H_{30}O$, molecular weight: 286.4516. The chemical structure is

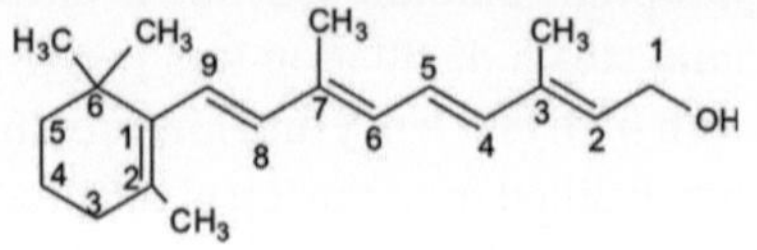

4.9.1 Properties of vitamin A

Vitamin A is a yellow flaky crystal or crystalline powder with a melting point of 62–64 °C and a boiling point of 137–138 °C (10^{-6} mmHg). It is a lipid-soluble vitamin, which is soluble in most organic solvents such as alcohols, ethers, hydrocarbons and halogenated hydrocarbons but insoluble in water and glycerin. The physical properties of vitamin A and vitamin A acetate are summarized in Table 4.9.

Table 4.9: Properties of vitamin A and vitamin A acetate.

Property	Vitamin A	Vitamin A acetate
Appearance	Yellow crystal	Yellow crystal
Odor	Faint, hay-like	Faint, hay-like
Molecular formula	$C_{20}H_{30}O$	$C_{22}H_{32}O_2$
Molecular weight	286.46	328.49

Table 4.9 (continued)

Property	Vitamin A	Vitamin A acetate
Melting point, °C	62–64	57–59
Solubility, g 100 mL^{-1}		
Water	Insoluble	Insoluble
Ethanol	Soluble	Soluble
Isopropanol	Soluble	Soluble
Chloroform	Soluble	Soluble
Acetone	Soluble	Soluble
Fats, oils	750	750
Spectrophotometric property, nm, max	375	326
Fluorescence		
Excitation, nm, max	325	325
Emission, nm, max	470	470

There are four conjugated double bonds in the side chain of vitamin A; thus, 16 geometric isomers in theory are possible. Due to the steric hindrance effect, the only geometric isomers existing in nature are the unhindered all-*E*, 9*Z*-, 13*Z*-, 9*Z*-, 13*Z*- and hindered 11*Z*-vitamin A. Only the all-*E* isomer has full activity of vitamin A. Vitamin A is unstable. It is easily destroyed by acid, air, oxidizing substances, high temperature and ultraviolet radiation. Since vitamin A acetate is more stable than vitamin A alcohol, the commercial form of vitamin A is mainly vitamin A acetate. Usually, the commercial product called vitamin A on the market is actually the acetate of vitamin A.

4.9.2 Production processes of vitamin A

In the early years before the advent of an industrial synthetic method, vitamin A was produced by the extraction from various marine animals but the concentration of vitamin A was very different in each animal and the source is limited. Chemical synthesis of vitamin A in laboratory began in 1937 but it was not until 1947 that Hoffmann-La Roche developed the first industrial synthesis method. Since then, vitamin A is mainly produced by chemical synthesis in industry. Vitamin A acetate is the main commercial form of vitamin A. At present, vitamin A acetate is chiefly manufactured by BASF, DSM (former vitamin division of Hoffmann-La Roche), Adisseo (former vitamin division of Rhône-Poulenc), Xiamen Kingdomway and Zhejiang NHU. All of the manufacturers use β-ionone as the key intermediate. Up to now, four industrial processes have been developed and commercialized for the production of vitamin A acetate: by Roche route using Grignard reaction with C6 alcohol, BASF route involving a Wittig reaction with an acetate of C5 aldehyde, Rhône-Poulenc route using a Julia reaction with C5 chloroacetyl and Wittig–Horner route applying phosphonate as an intermediate. Among them, Roche route and BASF route dominate majority of the total production output all over the world [26].

4.9.2.1 Roche synthesis process ($C_{14} + C_6$ route)

In 1947, the research group led by O. Isler at Hoffmann-La Roche in Switzerland realized the total synthesis of vitamin A acetate, and in 1948, Roche was the first in the world to achieve industrial production. In Roche synthesis process, β-ionone (C14) and 3-methyl-2-penten-4-yn-1-ol (C6 alcohol) are used as starting materials, so it is also called C14 + C6 route [27–29].

Figure 4.9 shows the Roche route to vitamin A acetate. β-Ionone undergoes a Darzen's condensation with methyl chloroacetate to produce the C14 aldehyde. The side chain is constructed by a Grignard reaction with C6 alcohol, resulting in the generation of diol, which is partially hydrogenated and acylated to obtain a monoacetate. Elimination and isomerization of the monoacetate give vitamin A acetate.

Some steps in the Roche synthesis process have been improved. For example, the partial hydrogenation was improved by Lindlar who developed a lead-doped $Pd/CaCO_3$ catalyst with quinoline poison, increasing the selectivity to the desired product. The elimination and isomerization of the monoacetate were also improved by using phosphorus oxychloride or a strong acid at low temperature.

The advantages of the Roche synthesis process include relatively mature technology, mild reaction conditions, stable yield, determined the stereoconfiguration of each reaction intermediate and no requirement for special raw materials. However, the synthesis route is very long, and more than 40 kinds of raw and auxiliary materials are used, resulting in a complicated management and control of the production. Nevertheless, the Roche synthesis process is widely used in the production of vitamin A acetate in industry, and manufacturers who use this process to produce vitamin A acetate mainly include DSM, Xiamen Kingdomway and Zhejiang NHU.

4.9.2.2 BASF synthesis process (C15 + C5)

BASF began to develop a different route for synthesis of vitamin A in the 1950s. After more than 10 years of continuous improvements, BASF started the industrial production of vitamin A acetate in 1971. A typical feature of this process is C15 + C5 route involving a Witting reaction. The reaction process is shown in Figure 4.10. β-Ionone reacts with acetylene to form ethynyl-β-ionol, followed by partial hydrogenation to obtain vinyl-β-ionol. Alternatively, ethynyl-β-ionol could be obtained by direct reaction of β-ionone with a vinyl Grignard. The vinyl-β-ionol was transformed into a C15 phosphonium salt by reaction with triphenylphosphine and HCl. The resulting C15 phosphonium salt underwent a Witting reaction with C5 aldehyde in the presence of sodium methoxide catalyst to produce vitamin A acetate [30].

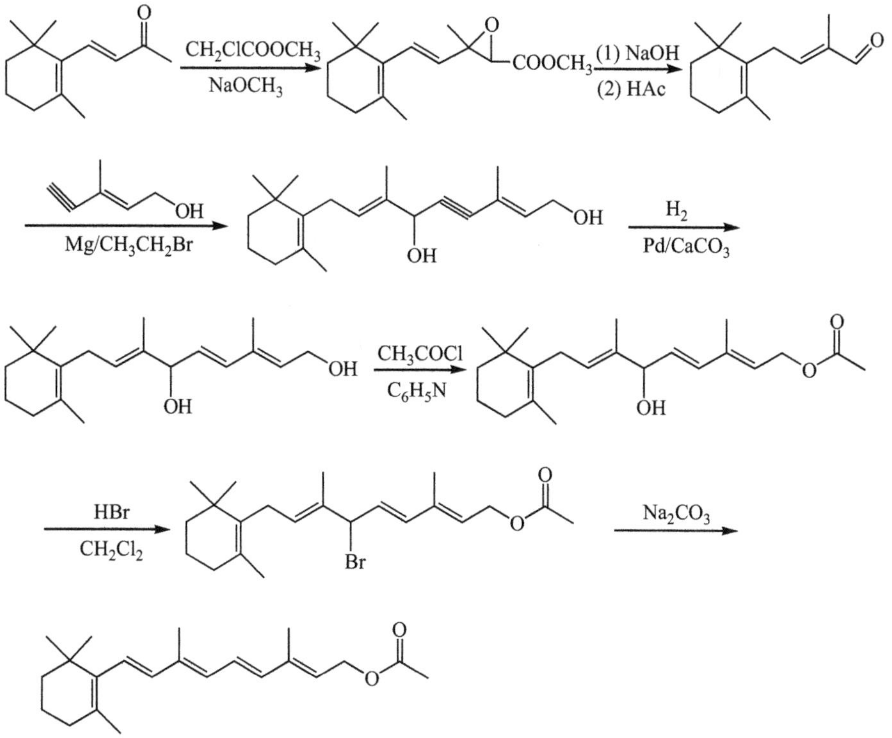

Figure 4.9: Roche route (C14 + C6) to vitamin A acetate.

4.9.2.3 Rhône-Poulenc route (C15 + C5) to vitamin A acetate

Rhône-Poulenc developed another C15 + C5 route using sulfone as an intermediate. Figure 4.11 represents the reaction process for vitamin A acetate by Rhône-Poulenc synthesis (C15 + C5). Vinyl-β-ionol was obtained by ethynylation and partial hydrogenation of β-ionone in the same way as BASF route. Vinyl-β-ionol reacted with sodium benzene sulfinate to produce an allylic sulfone (C15). The sulfone underwent a Julia reaction with C5 chloroacetyl to generate a C20 sulfone, followed by an elimination reaction to produce vitamin A acetate [26].

4.9.2.4 Wittig–Horner route (C15 + C5) to vitamin A acetate

Zhejiang Medicine Co. Ltd. has developed the Wittig–Horner route to vitamin A acetate using a phosphonate as an intermediate. The reaction process is shown in Figure 4.12. The C14 aldehyde was prepared by that same method as Roche route, and reacted with tetraethyl methylene disphosphonate in strong alkaline environment to obtain a diethyl phosphonate of 3-methyl-5-(2,6,6-trimethyl-1-cyclohexene)-1,4-pentadiene (C15). Diethyl C15 phosphonate reacted with C5 aldehyde in a strongly alkaline environment to give vitamin A acetate [26].

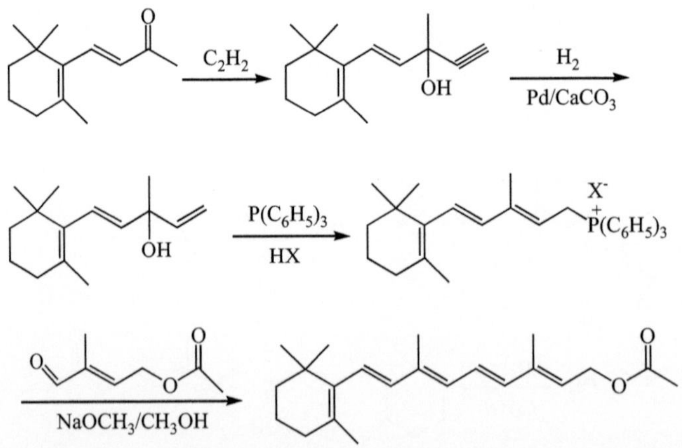

Figure 4.10: BASF route (C15 + C5) to vitamin A acetate by using a Wittig reaction.

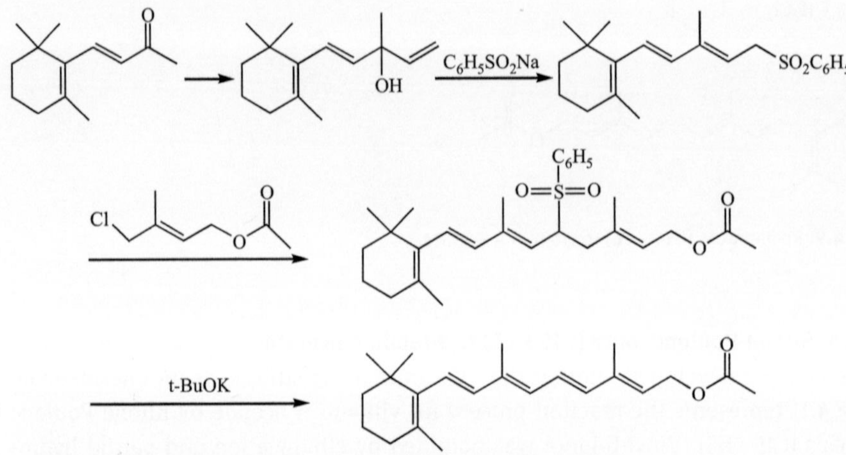

Figure 4.11: Rhône-Poulenc synthesis (C15 + C5) of vitamin A acetate.

4.9.3 Uses of vitamin A

Vitamin A is applied as a dietary supplement and an important medicine for the treatment of deficiency syndromes. It is not an endogenously generated vitamin; therefore, it is necessary to be supplied from dietary or vitamin supplement sources. The physiological function of vitamin A is to maintain the integrity of epithelial tissue and the permeability of cell membrane. It is very important to maintain normal vision. Moreover, it is also helpful for reproduction and lactation, and for promoting the growth and development of the human body and animals. In vitamin A-deficient human body and animals, epithelial changes such as keratinization, dry, scaly skin,

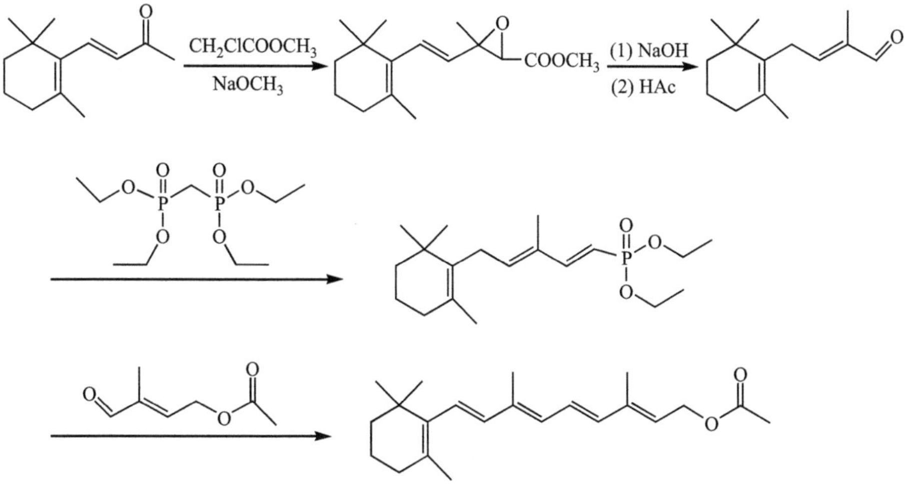

Figure 4.12: Wittig–Horner route (C15 + C5) to vitamin A acetate.

changes in the epidermis of the respiratory tract and susceptibility to bacterial infections can occur. There will also be insufficient rhodopsin, weak adaptability to dark light, night blindness and conjunctivitis.

References

[1] Wildermann A, Bonrath W, Hellmann G Transport of ethyne in form of α-alkynols as ethyne precursors. US 8003843 B2, 2011.

[2] Werner B, Peter S, Johannes T, et al. Ethynylation process. WO2004018400A1, 2004.

[3] Zhong C-R, Lu A, Rao Z-K. Synthesis and corrosion inhibitory property of methylpentynol, Oilfield Chemistry 1999, 16(2), 121–124.

[4] Hu W. Separation of tert-pentanol and water mixture by means of combination of liquid-liquid extraction rectification, Natural Gas Chemical Industry 1995, 20(4), 33–38.

[5] Kindler A, Brunner M, Tragut C Method for producing alkyne diols. US 6297407, 2001.

[6] Liu X. Development of reaction engineering of 2,5-dimethylhexynediol, Natural Gas Chemical Industry 1996, 21(6), 36–40.

[7] Wang J. Synthesis technology of 2,5-dimethylhexynediol, Natural Gas Chemical Industry 1985, 6, 59–56.

[8] Kimel W, Sax N W, Kaiser S. Total synthesis of pseudoionone and an isomeric ketone, Journal of Organic Chemistry 1958, 23(2), 153–157.

[9] Pasedach H, Mohr R, Nissen A Manufacture of 2-methyl-2-hepten-6-one. US 4173588, 1979.

[10] Sun Y, Zhang D, Zhao Z. Preparation of hydroxycitronellol from isobutylene, acetone and formaldehyde, Petrochemical Technology 1985, 15(6), 372–376.

[11] Weng Y, Guo S, Cai R. Synthesis of dehydrolinalool by continuous ethynylation process under pressure, Fine Chemicals 2002, 19(5), 297–299.

[12] Mourier E Process for producing acetylenic alcohols. 3,700,740, 1972.

[13] Shi J H, Zhang Q, Shi C. Kinetics of catalytic hydrogenation of dehydrolinalool on Raney-Ni catalyst, Journal of Zhejiang University of Technology 2006, 34(6), 606–610.
[14] Lapczynski A, Bhatia S P, Letizia C S, Api A M. Fragrance material review on dehydrolinalool, Food and Chemical Toxicology 2008, 46(11), S117-S120.
[15] Api A M, Belsito D, Bhatia S, et al. RIFM fragrance ingredient safety assessment, Linalool, CAS registry, Food and Chemical Toxicology 2015, 82, S29–S38.
[16] Yang H, Wu Z, Li X, et al. Synthesis of geranyl acetone from linalool catalyzed by sodium dihydrophosphate, China Surfactant Detergent & Cosmetic 2010, 40(4), 266–269.
[17] McGinty D, Letizia C S, Api A M. Addendum to fragrance material review on nerolidol (isomer unspecified), Food and Chemical Toxicology 2010, 48, S43–S45.
[18] Xie J, Ma H, Liu H, et al. Improved synthesis of isophytol catalyzed by aluminium isopropoxide, Chemical Research and Application 2015, 27(5), 629–634.
[19] Kido Y, Kumagai N, Iwasaki H Process for producing 6-methyl-3-hepten-2-one and 6-methyl-2-heptanone analogues, and process for producing phytone or isophytol. US 6232506 B1, 2001.
[20] McGinty D, Letizia C S, Api A M. Fragrance material review on isophytol, Food and Chemical Toxicology 2010, 48(3), 76–81.
[21] Chernysh G P, Koreshkova E G, Mel'nikova A A, Samokhvalov G I. Kinetics of the allylic rearrangement of 3-methylpent-1-en-4-yn-3-ol, Pharmaceutical Chemistry Journal 1974, 8(11), 688–690.
[22] Staroverov V M, Sleta L A, Drushlyak T G, Chernysh G P. Mechanism of allylic rearrangement of 3-methyl-1-penten-4-yn-3-ol, Pharmaceutical Chemistry Journal 1981, 15(9), 662–666.
[23] Wang G, Tian J, Liu Z. Study on catalysis of 3-methyl-1-penten-4-yn-3-ol with new acidic ionic liquid, Zhejiang Chemical Industry 2016, 47(8), 42–44.
[24] Pan B-W. Continuous steam distillation of 3-methyl-2-penten-4-yn-1-ol under reduced pressure, Huaxue Gongcheng & Zhuang Bei 2015, 1, 24–26.
[25] Gloor A, Gruendler H, Simon W Catalytic isomerization of Z-pentol to E-pentol. US 6465698 B2, 2002.
[26] Eggersdorfer M, Laudert D, Létinois U, et al. One hundred years of vitamins – a success story of the natural sciences, Angewandte Chemie International Edition 2012, 51, 12960–12990.
[27] Isler O. History and industrial application of carotenoids and vitamin A, Pure and Applied Chemistry 1979, 51, 447–462.
[28] Mercier C, Chabardes P. Organometallic chemistry in industrial vitamin A and vitamin E synthesis, Pure & Applied Chemistry 1994, 66(7), 1509–1518.
[29] Paust J. Recent progress in commercial retinoids and carotenoids, Pure & Applied Chemistry 1991, 63(1), 45–58.
[30] Pommer H, Nürrenbach A. Industrial synthesis of terpene compounds, Pure and Applied Chemistry 1975, 43, 527–555.

Chapter 5
Derivatives from acetylene reacting with halogen

5.1 1,1,2,2-Tetrachloroethane

1,1,2,2-Tetrachloroethane, also known as *s*-tetrachloroethane, acetylene tetrachloride, tetrachloroethane and sym-tetrachloroethane, is a derivative of acetylene and chlorine. CAS Registry Number: 79-34-5. Molecular formula: $C_2H_2Cl_4$, molecular weight: 167.849. The chemical structure is

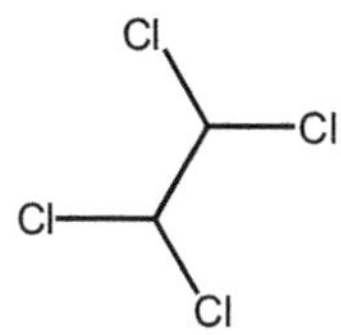

5.1.1 Properties of 1,1,2,2-tetrachloroethane

1,1,2,2-Tetrachloroethane is a non-flammable, non-explosive, colorless and transparent liquid with a strong odor similar to chloroform. The boiling point is 146.3 °C, and the melting point is −42.5 °C. It can be miscible with most organic solvents such as alcohol, ether, petroleum ether, halogenated hydrocarbon and carbon disulfide. It has the strongest dissolving power among chlorinated hydrocarbons. It can dissolve various organic substances including lipid, wax, asphalt, coal tar, camphor, rubber, dye, ethyl cellulose, nitrocellulose and polyvinyl chloride as well as inorganic substances such as sulfur, phosphorus, halogen and sodium sulfite. At 120 °C, 100 g of 1,1,2,2-tetrachloroethane can dissolve 100 g of sulfur. It is slightly soluble in water. The solubility in water at 25 °C is 0.29% (mass fraction), while the solubility of water in 1,1,2,2-tetrachloroethane is 0.13%. It forms an azeotrope with water, having an azeotropic point of 93.2 °C and an azeotropic composition: 1,1,2,2-tetrachloroethane of 68.9% and water of 31.1%. The physical properties of 1,1,2,2-tetrachloroethane are listed in Table 5.1.

The toxicity of this product is similar to chloroform, and the allowable concentration in air at working place is 5×10^{-6} in China.

The impure product of 1,1,2,2-tetrachloroethane is easily decomposed into hydrogen chloride and trichloroethylene, while pure product is relatively stable in the absence of air, moisture and light, but are easily decomposed into trichloroethylene in the presence of alkali. When 1,1,2,2-tetrachloroethane is in contact with air, it slowly removes hydrogen chloride to form trichloroethylene and traces of phosgene. In the presence of moisture, it gradually decomposes and releases hydrogen chloride.

https://doi.org/10.1515/9783110714999-005

Table 5.1: Physical properties of 1,1,2,2-tetrachloroethane.

Melting point/°C	−42.5	Enthalpy of formation of gas (25 °C)/kJ · mol^{-1}	149.0
Boiling point/°C	146.1	Enthalpy of combustion/kJ · kg^{-1}	5,788.8
Density (20 °C)/g · cm^{-3}	1.5953	Enthalpy of vaporization/kJ · mol^{-1}	37.64
Refractive index (20 °C)	1.4942	Enthalpy of fusion/kJ · mol^{-1}	9.17
Dipole moment/(×10^{-30} C · m)	5.70	Heat capacity (20 °C)/kJ · kg^{-1} · K^{-1}	1.122
Dielectric constant (20 °C)	8.0	Critical temperature/°C	371.3
Viscosity (20 °C)/mPa · s	1.77	Thermal conductivity/W · m^{-1} · K^{-1}	0.1364
Surface Tension (20 °C)/N · m^{-1}	36.04 × 10^{-3}	Electric conductivity/Ω^{-1} · cm^{-1}	4.5 × 10^{-9}

1,1,2,2-Tetrachloroethane is converted into dichloroacetyl chloride by irradiation with ultraviolet light in the presence of air or oxygen. It does not react with chlorine at room temperature, but chlorination occurs under ultraviolet light, resulting in the generation of hexachloroethane. 1,1,2,2-Tetrachloroethane is reduced to 1,2-dichloroethylene by treating with a metal such as iron, aluminum or zinc in the presence of boiling water or steam. It generates a highly explosive dichloroacetylene upon heating in the presence of a strong base.

5.1.2 Process for manufacture of 1,1,2,2-tetrachloroethane

1,1,2,2-Tetrachloroethane is prepared by addition reaction of acetylene with chlorine:

$$HC{\equiv}CH + 2Cl_2 \longrightarrow CHCl_2CHCl_2$$

Since direct reaction of the gaseous acetylene with chlorine may cause an explosion, the reaction is carried out using 1,1,2,2-tetrachloroethane itself as the solvent. The catalyst is stibium pentachloride or ferric chloride. When the ferric chloride catalyst is used, dry acetylene and chlorine are continuously introduced into the tetrachloroethane which is kept in reflux under reduced pressure. The heat of reaction is absorbed by evaporation of the tetrachloroethane and removed by the reflux condenser. The yield of 1,1,2,2-tetrachloroethane is 97% based on acetylene.

In the process of chlorination of acetylene to tetrachloroethane, the degree of vacuum of the chlorination tower directly affects the reaction temperature, which is the determining factor of the composition of the product and the reaction rate; meanwhile, maintaining a stable vacuum plays an important role in the safe operation of the production system. In general, the vacuum level in the chlorination tower is controlled to be about −90 kPa, which is very suitable for the production of tetrachloroethane by chlorination of acetylene. Under this vacuum, the reaction temperature is about 90 °C. The operation flexibility of the chlorination tower is large under higher vacuum, which is beneficial to the smooth and safe operation of the chlorination tower.

5.1.3 Uses of 1,1,2,2-tetrachloroethane

1,1,2,2-Tetrachloroethane is used as an intermediate for trichloroethylene, tetrachloroethylene, pentachloroethane and hexachloroethane. As a nonflammable solvent, it can be used to dissolve shellac, resins, waxes, and others but it is highly toxic and prone to hydrolysis, which limits its application. In addition, it is also used in the production of metal detergents, paint removers, insecticides, herbicides, alcohol denaturants and the like.

5.2 1,1,2,2-Tetrabromoethane

1,1,2,2-Tetrabromoethane, also known as acetylene tetrabromide, sym-tetrabromoethane, tetrabromoacetylene and *s*-tetrabromoethane, is a derivative of acetylene and bromine. CAS Registry Number: 79-27-6. Molecular formula: $C_2H_2Br_4$, molecular weight: 345.653. The chemical structure is

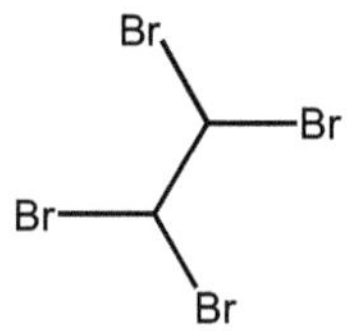

5.2.1 Properties of 1,1,2,2-tetrabromoethane

1,1,2,2-Tetrabromoethane is a colorless to light yellow, transparent and oily liquid with a boiling point of 243.5 °C. The industrial product is a light yellow or amber liquid with an odor of camphor and iodoform. It is miscible with organic solvents such as alcohol, ether, chloroform, carbon tetrachloride, aniline and glacial acetic acid. 1,1,2,2-Tetrabromoethane is slightly soluble in water with a solubility of 0.065% at 30 °C. The physical properties of 1,1,2,2-tetrabromoethane are listed in Table 5.2.

1,1,2,2-Tetrabromoethane reacts with a strong base to generate hydrogen bromide. It is stable at room temperature, yet it will be decomposed upon heating to above 240 °C, resulting in the release of bromine and hydrogen bromide.

5.2.2 Process for manufacture of 1,1,2,2-tetrabromoethane

1,1,2,2-Tetrabromoethane is obtained by addition reaction of acetylene with bromine [1–4]. The reaction proceeds easily in the absence of a catalyst. In order to avoid the

Table 5.2: Physical properties of 1,1,2,2-tetrabromoethane.

Melting point/°C	−1.0	Surface tension/N · m^{-1}	
Boiling point/°C	243.5	25 °C	49.07 × 10^{-3}
Vapor pressure/kPa		110 °C	1.33
125 °C	2.0	Enthalpy of fusion/kJ · kg^{-1}	33.78
65 °C	0.13	Enthalpy of vaporization (428 K)/ kJ · mol^{-1}	56.9
Density (20 °C)/g · cm^{-3}	2.9501	Heat capacity (25 °C)/J · mol^{-1} · K^{-1}	165.7
Refractive index (20 °C)	1.6353	Critical temperature/°C	388
Dipole moment(×10^{-30} C · m)	4.37	Thermal conductivity/W · m^{-1} · K^{-1}	0.1364
Dielectric constant (25 °C)	6.60	Electric conductivity/Ω^{-1} · cm^{-1}	4.5 × 10^{-9}
Viscosity (20 °C)/mPa · s	9.80		

occurrence of substitution reactions and other side reactions, the reaction is usually carried out at a lower temperature (<60 °C). The reaction equation is as follows:

$$HC{\equiv}CH + 2Br_2 \longrightarrow CHBr_2CHBr_2$$

Bromine was added to the glass-lined reactor and a small amount of water was added to block the bromine liquid surface to prevent it from volatilization. Then, acetylene was slowly introduced into the reactor with a pressure of 0.02–0.05 MPa. The reaction was carried out at a temperature of 50–60 °C for 5–7 h. During the reaction, the liquid gradually turned yellow. The reaction solution was washed with a dilute alkali solution, and then allowed to stand for separating into two layers. The lower organic phase was collected, washed with water until neutral to obtain 1,1,2,2-tetrabromoethane.

The 1,1,2,2-tetrabromoethane synthesis reaction is performed in two steps. First, acetylene reacts with bromine to form dibromoethylene, and then the resultant dibromoethylene further reacts with bromine to give 1,1,2,2-tetrabromoethane:

$$C_2H_2 \xrightarrow{Br_2} CHBr{=}CHBr \xrightarrow{Br_2} CHBr_2 - CHBr_2$$

Both alkynes and alkenes can undergo an electrophilic addition reaction. Since there are two π bonds in the acetylene triple bond, they can generally be added with two molecular reagents. Apparently, it seems that the alkyne is easier to proceed electrophilic addition than the alkene, but in fact, bromine preferentially undergoes an addition reaction with the alkene to form tetrabromoethane. Therefore, side reactions are not easy to occur. However, if the acetylene is largely excessive, by-products such as dibromoethylene and tribromoethane are generated.

The composition of reaction solution produced from the synthesis tower is relatively complicated. If the actual ratio of bromine and acetylene is closer to the stoichiometric ratio during the reaction, the impurity content in the reaction solution is less, and content of 1,1,2,2-tetrabromoethane is higher. In practice, acetylene is only slightly excessive, for example, 1–5% excess. If the excess of acetylene is more than 10%, the content of dibromoethylene and tribromoethane in the reaction solution is significant, as shown in Table 5.3.

Table 5.3: Contents of each component of the synthetic liquid when acetylene is excessive.

Component	Content, %	
	Acetylene 5% excess	Acetylene 10% excess
$C_2H_2Br_2$	2.42	17.70
$CHBr_3$	0.31	2.55
$C_2H_3Br_3$	1.31	5.18
$C_2H_2Br_4$	95.97	72.32

Increasing the illumination (visible or ultraviolet light) can promote the addition reaction of acetylene and bromine, shorten the reaction time, increase the content of 1,1,2,2-tetrabromoethane, and reduce the content of by-products in the reaction solution.

A pH of 5 to 7 is required for the finished product of 1,1,2,2-tetrabromoethane, which is close to neutral. However, hydrobromic acid is present throughout the production process, resulting in an unacceptable acidity index. Usually in industry, solid sodium carbonate is added to the acidic reaction solution to neutralize the acid produced during the synthesis. The base added here must be a solid rather than an aqueous solution; otherwise, a side reaction of elimination will occur.

5.2.3 Uses of 1,1,2,2-tetrabromoethane

1,1,2,2-Tetrabromoethane is used as a cocatalyst for the oxidation process in the production of terephthalic acid. It is also used as a flame retardant in plastic products such as expandable polystyrene foam. Furthermore, it is an intermediate for synthesis of quaternary ammonium salts, fuel, refrigerant, disinfectant and pesticides. Moreover, it is applied as a mineral flotation agent and oil field killing fluid.

5.2.4 Status of production and market of 1,1,2,2-tetrabromoethane in China

In China, more than 10 manufacturers are engaged in the production of 1,1,2,2-tetrabromoethane. Among them, Tianjin Changlu Haijing Group Co., Ltd. has the largest production output. The product is mainly provided to Shanghai Petrochemical Complex, Nanjing Yangzi, Yanshan Petrochemical, Guangdong Maoming Petrochemical, etc. Jiangsu Dacheng Pharmaceutical Technology Co., Ltd. has a production capacity of 300 t a^{-1}. Other manufacturers of 1,1,2,2-tetrabromoethane are located in Shandong, Jiangsu and Zhejiang, but the production outputs are small.

The market demand for 1,1,2,2-tetrabromoethane is more than 1,000 t a^{-1} in China. Some newly built petrochemical plants are temporarily transitioning with imported products. Some petrochemical companies use hydrobromic acid instead of tetrabromoethane as a cocatalyst.

5.3 Trichloroethylene

Trichloroethylene, also known as chlorylene, 1,1,2-trichloroethylene, 1,1,2-trichloroethene, acetylene trichloride, 1-chloro-2,2-dichloroethylene, 1,1-dichloro-2-chloroethylene, trichlorethene and 1,2,2-trichloroethylene, is a derivative of acetylene and chlorine. CAS Registry Number: 79-01-6. Molecular formula: C_2HCl_3, molecular weight: 131.388. The chemical structure is

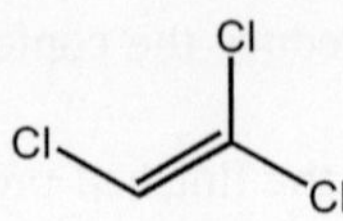

5.3.1 Properties of trichloroethylene

Trichloroethylene is a colorless and transparent liquid with an odor similar to chloroform. It is volatile and nonflammable. Trichloroethylene is insoluble in water. It is soluble in most organic solvents such as ethanol and ether. Trichloroethylene forms an azeotrope with water, having an azeotropic point of 73.6 °C and azeotropic composition: trichloroethylene of 94.6% and water of 5.4%. Trichloroethylene has an anesthetic effect. It can cause anesthesia after 20 min in the air containing 400×10^{-6} of trichloroethylene. It will lose consciousness after 10 min in the air containing $3{,}000 \times 10^{-6}$ of trichloroethylene. The physical properties of trichloroethylene are shown in Table 5.4.

Usually, a stabilizer is added to the trichloroethylene during storage to prevent it from auto-oxidation. The trichloroethylene containing no stabilizer is gradually oxidized in the air to form an intermediate dichloroacetyl chloride and subsequently generate phosgene, carbon monoxide and hydrogen chloride:

$$ClCH{=}CCl_2 + O_2 \longrightarrow Cl_2CHCOCl \longrightarrow HCl + CO + COCl_2$$

The reaction proceeds according to the radical mechanism, and light and heat significantly accelerate the reaction. In the presence of water, dichloroacetyl chloride is

Table 5.4: Physical properties of trichloroethylene.

Melting point/°C	−86.4	Enthalpy of formation of gas (25 °C)/kJ · mol^{-1}	5.86
Boiling point/°C	87.2	Enthalpy of combustion/kJ · mol^{-1}	963.0
Density (20 °C)/g · cm^{-3}	1.4649	Enthalpy of vaporization (360.4 K)/kJ · mol^{-1}	31.4
Refractive index (20 °C)	1.4782	Enthalpy of fusion (188.5 K)/ kJ · mol^{-1}	8.45
Dipole moment/(×10^{-30} C · m)	3.0	Heat capacity (20 °C)/kJ · kg^{-1} · K^{-1}	0.93
Dielectric constant (20 °C)	3.409	Critical temperature/°C	298
Viscosity (20 °C)/mPa · s	0.58	Critical pressure/MPa	4.92
Flash point/°C	no	Thermal conductivity/W · m^{-1} · K^{-1}	0.1386
Ignition point/°C	425	Electric conductivity/Ω^{-1} · cm^{-1}	8 × 10^{-12}

decomposed into dichloroacetic acid and hydrogen chloride. The acid produced by the decomposition corrodes the metal. Therefore, a trace amount of a stabilizer such as hydroquinone or an amine is added to it. In the presence of air, moisture and light, even if it is heated to 130 °C, the trichloroethylene with addition of a stabilizer has no effect on metal materials generally used in industry. When trichloroethylene vapor is heated to more than 700 °C, it decomposes into a mixture of dichloroethylene, tetrachloroethylene, carbon tetrachloride and chloroform.

Trichloroethylene is not easily hydrolyzed under normal conditions of use, but it can be hydrolyzed to chloroacetic acid by heating to 90 °C in the presence of 90% sulfuric acid:

$$ClCH{=}CCl_2 + H_2O \longrightarrow ClCH_2COOH + HCl$$

In the presence of a copper salt, trichloroethylene is heated to 175 °C under pressure and reacts with an aqueous solution or suspension of an alkali metal or alkaline earth metal hydroxide to form a hydroxyethyl salt. Trichloroethylene does not react with hydrochloric acid or nitric acid at low temperature. However, upon heating, trichloroethylene reacts violently with concentrated nitric acid and decomposes completely into trichloronitromethane or monochlorodinitromethane according to the reaction conditions. In the presence of aluminum trichloride, trichloroethylene reacts with hydrogen chloride to form 1,1,1,2-tetrachloroethane, with chloroform to form 1,1,1,2,3,3-hexachloropropane, and with carbon tetrachloride to form 1,1,1,2,3,3,3-heptachloropropane.

Aluminum, especially aluminum powders can cause the decomposition of the trichloroethylene without a stabilizer to generate hydrogen chloride, and at the same time, a strong explosion occurs.

5.3.2 Production process of trichloroethylene

The industrial processes for production of trichloroethylene include three routes: chlorination of acetylene, direct chlorination of ethylene, and oxychlorination of ethylene. Before the 1970s, the production of trichloroethylene by chlorination of acetylene dominated, but since the 1980s, due to the rapid development of petrochemical industry, ethylene direct chlorination and ethylene oxychlorination have gradually become mainstream, but in China, trichloroethylene is mainly produced by chlorination of acetylene [5].

5.3.2.1 Chlorination of acetylene

Acetylene reacts with chlorine to form tetrachloroethane, and then the resultant tetrachloroethane is saponified to produce trichloroethylene. The reaction equations are as follows:

$$HC{\equiv}CH + 2Cl_2 \longrightarrow CHCl_2CHCl_2$$

$$2CHCl_2CHCl_2 + Ca(OH)_2 \longrightarrow 2CHCl{=}CCl_2 + CaCl_2 + 2H_2O$$

Acetylene and chlorine are introduced into a chlorination reaction tower containing a tetrachloroethane solvent and ferric chloride or antimony trichloride catalyst to start the reaction at a temperature of 70–90 °C and under a pressure of 9.8–29.4 kPa. The molar ratio of chlorine to acetylene is 2:2.05, and the dosage of ferric chloride catalyst is 0.1%. The heat of reaction is removed by the circulation of tetrachloroethane to an external heat exchanger. The obtained tetrachloroethane is introduced into an intermediate storage tank and sent to the trichloroethylene reaction tower for a saponification reaction with 10% lime milk at a temperature between 70 and 100 °C. Hydrogen chloride is removed from tetrachloroethane to give trichloroethylene. Steam is introduced into the bottom of the tower to boil the reaction solution. The azeotrope of trichloroethylene and water is distilled off from the top of the tower, condensed and separated into two layers. The organic layer is collected and subjected to fractional distillation towers for removal of the light and heavy components. The finished product of trichloroethylene is obtained with a total yield of more than 95%. Production of 1 ton of trichloroethylene consumes 198 m^3 of acetylene, 1,350 kg of chlorine and 2.4 t of steam.

The production of trichloroethylene by chlorination of acetylene has advantages including simple process and low investment. However, a large amount of calcium chloride mud was produced, which polluted the environment. At present, China has developed a clean production process that does not require a saponification reaction, that is, trichloroethylene is obtained by catalytic dehydrochlorination of tetrachloroethane in the gas phase at 280 °C. Since no lime milk is used, no calcium chloride mud is produced and thus no environmental pollution occurs.

5.3.2.2 Direct ethylene chlorination

Ethylene is directly chlorinated to obtain a mixture of tetrachloroethane and pentachloroethane, which are subjected to gas phase cracking to obtain trichloroethylene, and by-product tetrachloroethylene. The reaction equations are as follows:

$$CH_2CH_2 + Cl_2 \longrightarrow CHCl_2CHCl_2 + CHCl_2CCl_3 + HCl$$

$$CHCl_2CHCl_2 \longrightarrow ClCH{=}CCl_2 + HCl$$

$$CHCl_2CCl_3 \longrightarrow Cl_2C{=}CCl_2 + HCl$$

Ethylene and chlorine were introduced into the reaction tower, and the ratio of the raw materials could be adjusted according to the desired product ratio. The conversion of ethylene was 95% to 98%. The reaction solution was sent to a deacidification tower to remove dissolved hydrogen chloride. Subsequently, the solution was introduced into a fractional distillation tower. The top product was sent back to the reaction tower. The bottom product (a mixture of tetrachloroethane and pentachloroethane) was introduced into a cracking furnace for splitting into trichloroethylene and tetrachloroethylene at a temperature of 400 to 500 °C in the presence of a barium chloride catalyst supported on activated carbon. The outlet gas from the cracking furnace was quenched and separated from hydrogen chloride. Then, the resulting solution was introduced into fractional distillation towers for removal of light and heavy components, and finally refined trichloroethylene and tetrachloroethylene were obtained.

In this process, the production of trichloroethylene is always accompanied by the by-product tetrachloroethylene. The ratio of the two products can be adjusted by changing the ratio of raw materials.

5.3.2.3 Oxychlorination of ethylene

Ethylene undergoes an oxychlorination with chlorine and oxygen or air to obtain a trichloroethylene and tetrachloroethylene:

$$CH_2CH_2 + Cl_2 \longrightarrow CH_2ClCH_2Cl$$

$$8CH_2ClCH_2Cl + 6Cl_2 + 7O_2 \longrightarrow 4ClCH{=}CCl_2 + 4Cl_2C{=}CCl_2 + 14H_2O$$

Using copper chloride or potassium chloride as a catalyst, 85–90% of ethylene was converted into chlorinated hydrocarbons, and 10–15% was converted into carbon monoxide and carbon dioxide in a bubbling bed reactor at a temperature of 425 °C. At a higher reaction temperature, the side reaction such as decomposition was significant; and at lower reaction temperature, the generation of tetrachloroethane by-product was increased.

5.3.3 Uses of trichloroethylene

Trichloroethylene is an excellent solvent that can be used as a substitute for benzene and gasoline. It can be used to dissolve resins, asphalt, coal tar, cellulose acetate, nitrocellulose, rubber and coatings. Trichloroethylene is applied as a dry cleaning agent for textiles and wool, a surface treatment agent for metals, a cleaning agent for electronic components. It is also used as a raw material for production of tetrachloroethylene. In recent years, trichloroethylene has been used in large quantities to produce dichloroacetyl chloride, octachlorodipropyl ether, new refrigerants HFC-134a, HFC-123 and HFC-125. In addition, it is also used as caprolactam, pesticide extractant, general anesthetic, monochloroacetic acid and other organic synthesis intermediates.

5.3.4 Production status of trichloroethylene

Trichloroethylene, tetrachloroethylene and 1,1,1-trichloroethane are known as the three major organochlorine solvents. Trichloroethylene was developed in the 1920s and large-scale industrial production began in the early 1930s. In 2012, the total production capacity of trichloroethylene outside China was about 500,000 t a^{-1}, of which the United States had much more production capacity (PPG company 91,000 t a^{-1}, Dow Chemical company 55,000 t a^{-1}) than other countries.

In China, Fang Da Jinhua first used acetylene as a raw material to produce trichloroethylene through chlorination and saponification in the early 1970s. In recent years, the scale of chlor-alkali enterprises has expanded rapidly, and the abundant production of chlorine gas needs to be consumed, which has promoted the development of trichloroethylene as a chlorine-consuming product. Especially with the development of new refrigerants R-134a and R-125, the production of trichloroethylene has been pushed to a climax. Since 2007, many enterprises have built trichloroethylene projects and applied it as the main product that consumes chlorine. At present, there are more than ten manufacturers of trichloroethylene in China, with a total production capacity of about 363,000 t a^{-1}, as shown in Table 5.5. As some are newly built to be put into production, and some companies have stopped production, the actual output of trichloroethylene was 260,000 tons in 2012 [6].

It is estimated that the demand for trichloroethylene used as a raw material for the production of new refrigerants is about 200,000 t a^{-1}, and the amount of trichloroethylene used for the production of cleaning agents is about 50,000 t a^{-1}. The demand for trichloroethylene in other industries is about 10,000–30,000 t a^{-1}, that is, the total demand for trichloroethylene is between 260,000 and 280,000 t a^{-1}. Compared with actual output, supply and demand are basically balanced. As some newly built installations are put into production, there will be oversupply and overcapacity in China.

Table 5.5: Main manufacturers and production capacity of trichloroethylene in China.

No	Manufacturers	Production capacity/(t a^{-1})
1	Shandong Binhua Dongrui Chemical Co., Ltd.	80,000
2	Shandong Xinlong Group Co., Ltd.	60,000
3	Alxa Dakang Fine Chemical Co., Ltd.	48,000
4	Zhejiang Juhua Co., Ltd.	40,000
5	Yibin Haifeng Herui Co., Ltd.	30,000
6	Chongqing Yingtianhui Chlor-Alkali Chemical Co., Ltd.	30,000
7	Inner Mongolia Sanlian Chemical Co., Ltd.	20,000
8	Fangda Jinhua Chemical Technology Co., Ltd.	20,000
9	Ningxia Kangmei Chemical Co., Ltd.	15,000
10	Guizhou Lantian Chemical Co., Ltd.	15,000
11	Jiangxi Guohong Chemical Co., Ltd.	5,000

5.4 Tetrachloroethylene

Tetrachloroethylene, also known as ethylene tetrachloride, perchloroethylene, perchlorethylene, tetrachlorethylene, tetrachloroethene, 1,1,2,2-tetrachloroethylene and 1,1,2,2-tetrachloroethene, is a derivative of trichloroethylene. CAS Registry Number: 127-18-4. Molecular formula: C_2Cl_4, molecular weight: 165.833. The chemical structure is

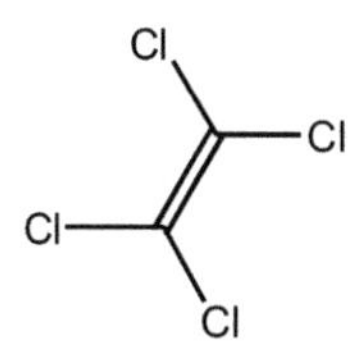

5.4.1 Properties of tetrachloroethylene

Tetrachloroethylene is a colorless liquid with an odor similar to ether. Its boiling point is 121.2 °C, and the melting point is −22.4 °C. It can be miscible with common organic solvents such as ethanol, ether, benzene, chloroform and carbon tetrachloride. It can dissolve fat, oil, tar, rubber, natural resin and aromatic organic acids. It is hardly soluble in water. The solubility of tetrachloroethylene in water is 0.015% (wt) at 25 °C; and the solubility of water in tetrachloroethylene is 0.0105% (wt). Tetrachloroethylene forms an azeotrope with water, having an azeotropic point of 87.7 °C and an azeotropic composition: tetrachloroethylene of 84.2% and water of 15.8%. The physical properties of tetrachloroethylene are listed in Table 5.6.

Tetrachloroethylene is an anesthetic that has a strong inhibitory effect on the central nervous system. Contact with liquid tetrachloroethylene causes skin and eyes

Table 5.6: Physical properties of tetrachloroethylene.

Melting point/°C	−22.4	Dielectric constant (25 °C)	2.3
Boiling point/°C	121.2	Viscosity (20 °C)/mPa · s	0.880
Vapor pressure/kPa		Surface tension (20 °C)/N · m^{-1}	32.32×10^{-3}
100.0 °C	58.46	Flash point/°C	non
70.0 °C	20.75	Enthalpy of formation of liquid (25 °C)/kJ · mol^{-1}	12.56
26.3 °C	2.67	Enthalpy of combustion/kJ · mol^{-1}	680.4
Density (15 °C)/g · cm^{-3}	1.6311	Enthalpy of vaporization (394.1 K)/kJ · mol^{-1}	34.68
Density (20 °C)/g · cm^{-3}	1.6226	Enthalpy of fusion (250.81 K)/ kJ · mol^{-1}	10.88
Refractive index (20 °C)	1.5057	Critical temperature/°C	347.1
Dipole moment/(×10^{-30} C · m)	0	Critical pressure/MPa	4.49

irritation. The vapor of tetrachloroethylene irritates the respiratory tract and causes central nervous system disease, which is life-threatening. In China, the maximum allowable concentration in the air of this product at working place is 100×10^{-6}.

In the presence of light, water and oxygen, tetrachloroethylene decomposes into acidic substances that are corrosive to metals. Usually, a derivative of an amine or phenol with a concentration between 10 and 500 mg/L is added to the tetrachloroethylene as a stabilizer or a decomposition inhibitor [7]. After the addition of the stabilizer, tetrachloroethylene has no significant corrosive effect on commonly used metal materials in the presence of air, moisture and light, even if it is heated to 140 °C. Tetrachloroethylene is also stable up to about 500 °C in the absence of catalyst, air and moisture. However, when passing through the red heat pipe together with the air, it decomposes into carbon monoxide, chlorine and phosgene by pyrolysis. Tetrachloroethylene is converted into hexachloroethane and hexachlorobenzene in contact with activated carbon at 700 °C. It reacts with ozone to form phosgene and trichloroacetyl chloride. It reacts with a mixture of sulfuric acid and nitric acid to form trichloroacetyl chloride and a small amount of tetrachlorodinitroethane. Tetrachloroethylene does not react with concentrated nitric acid upon heating, but reacts with fuming nitric acid to form trichloroacetyl chloride and tetrachlorodinitroethane. It reacts with nitrogen dioxide to form tetrachlorodinitroethane at a temperature of 100 °C. Hydrogenation of tetrachloroethylene yields tetrachloroethane.

5.4.2 Process for manufacture of tetrachloroethylene

The industrial processes for production of tetrachloroethylene include three technical routes: (1) chlorination of trichloroethylene and followed by dehydrochlorination; (2) oxychlorination of ethylene; and (3) chlorination and pyrolysis of light hydrocarbon. In China, the chlorination of trichloroethylene and dehydrochlorination process is mainly used for production of tetrachloroethylene, while the other two processes are widely used in other countries [5].

5.4.2.1 Chlorination of trichloroethylene and dehydrochlorination

The initial industrial production of tetrachloroethylene was achieved by photochlorination of acetylene with chlorine. Afterward, the process is improved. First, trichloroethylene is prepared from acetylene and chlorine; second, chlorination of trichloroethylene produces pentachloroethane; third, dehydrochlorination of pentachloroethane yields tetrachloroethylene. The whole process includes two chlorination reactions and two saponification reactions. The production of trichloroethylene has been described in Section 5.3. Only the subsequent steps from trichloroethylene are described here. The reaction equations are as follows:

$$2CHCl=CCl_2 + Cl_2 \longrightarrow CHCl_2CCl_3$$

$$2CHCl_2CCl_3 + Ca(OH_2) \longrightarrow 2Cl_2C=CCl_2 + CaCl_2 + 2H_2O$$

A certain amount of trichloroethylene (industrial first grade) and ferric chloride catalyst were added to the reaction tower and preheated to 50 °C. Chlorine gas was introduced into the reaction tower. When the reaction was intense, the heat released from the reaction is removed by cooling water, and the reaction temperature was maintained at 60 to 80 °C until the reaction reached the end point. The reaction solution and a certain amount of lime milk were added to the saponification kettle, and the saponification temperature was about 90 °C but controlled to be not more than 95 °C. The crude tetrachloroethylene produced after the saponification reaction is subjected to fractional distillation towers for removal of light and heavy components under vacuum; thereby, a finished product of tetrachloroethylene was obtained.

5.4.2.2 Oxychlorination of ethylene

This method is the same as the oxychlorination method in the production process of trichloroethylene, which has been described in Section 5.3.

5.4.2.3 Chlorination and pyrolysis of light hydrocarbon

Tetrachloroethylene is obtained by chlorination and pyrolysis of C1 to C3 hydrocarbons, and a by-product carbon tetrachloride is also produced. The reactions are as follows:

$$2CH_4 + 6Cl_2 \longrightarrow Cl_2C=CCl_2 + 8HCl$$

$$2CH_3CH_3 + 10Cl_2 \longrightarrow 2\,Cl_2C=CCl_2 + 12HCl$$

$$2CH_3CH_2CH_3 + 14Cl_2 \longrightarrow 3\,Cl_2C{=}CCl_2 + 16\,HCl$$

$$CH_3CH_2CH_3 + 10\,Cl_2 \longrightarrow 3\,CCl_4 + 8\,HCl$$

C1 to C3 hydrocarbons, chlorine gas and recycle gas were injected into the cracking reactor, and the reaction was carried out at a temperature of 550–700 °C under atmospheric pressure. No catalyst was required in the reaction. The reaction product was quenched with concentrated hydrochloric acid (21–36%) in a quench tank. The organic compounds in reaction mixture were condensed, and the hydrogen chloride was absorbed in a hydrochloric acid absorption tower with 21% hydrochloric acid or water. The condensate in the quench tank was sent to the separation tank for separating into two layers. The water layer was used as quenching hydrochloric acid. The excess hydrochloric acid was introduced into the chlorine desorption tower for removal of chlorine dissolved in it, and the resultant 35% hydrochloric acid was sent to hydrogen chloride distillation tower. Anhydrous hydrogen chloride gas was obtained from the top of the tower, and 21% hydrochloric acid was collected at the bottom of the tower and sent to the hydrochloric acid absorption tower. The organic layer of the separation tank was sent to another desorption tower to remove the chlorine and hydrogen chloride dissolved in it, and then introduced into fractional distillation system to obtain pure tetrachloroethylene and carbon tetrachloride. The bottom product of each distillation tower was separately recovered in an evaporator, and the heavy components were recycled to the reactor. The total yield of tetrachloroethylene and carbon tetrachloride was 96% based on chlorine.

This process was developed by Scientific Design Company (United States). Kureha Corporation (Japan) also adopted the process to produce tetrachloroethylene. The difference was that a dry quench system was used by Scientific Design Company, while hydrochloric acid was applied to quench the reaction mixture in Kureha Corporation. Since the reaction is a strong exothermic reaction, the rapid removal of a large amount of heat released by the reaction is the key to the smooth operation of the reactor. The recycle gas is a mixture of excess chlorine, lower hydrocarbon chloride and heavy components recovered from the evaporator.

The by-product carbon tetrachloride can be recycled to the reactor and converted into tetrachloroethylene, and vice versa. The ratio of tetrachloroethylene to carbon tetrachloride can be adjusted according to the market demand. The by-product hydrogen chloride was recovered as hydrochloric acid, or as a raw material for manufacture of vinyl chloride, thereby obtaining a high degree of comprehensive utilization.

5.4.3 Uses of tetrachloroethylene

Tetrachloroethylene is mainly used as a dry-cleaning agent. It is an intermediate for production of trichloroacetic acid and fluorine-containing organic compounds. It is also used as a solvent for many organic substances. In addition, it is used in metal degreasing.

5.4.4 Production and market status of tetrachloroethylene

The production of tetrachloroethylene is concentrated in the United States, European Union, China, Japan and Russia. In recent years, the production capacity of tetrachloroethylene in China increased rapidly, while other countries have little production growth.

The production of tetrachloroethylene in China started in the 1980s. By 2012, there were more than a dozen manufacturers of tetrachloroethylene with a scale of more than 10,000 t a^{-1}, which provided a total production capacity of 190,000 t a^{-1}, including Shandong Xinlong Technology (40,000 t a^{-1}) and Zhejiang Quzhou Giant Plastic Chemical (40,000 t a^{-1}). However, most companies were not operating at full capacity. In 2011, the actual product output of tetrachloroethylene in China was only 59,000 tons.

The consumption of tetrachloroethylene in China is as follows: degreasing cleaning and dry-cleaning accounts for 65%, intermediates for organic synthesis accounts for 20%, and other usages account for 15%. In 2013, the market consumption of tetrachloroethylene was in the range from 60,000 to 70,000 tons. According to customs statistics, the imports of tetrachloroethylene dropped from 15,000 tons in 2008 to 11,000 tons in 2012, and the export volume increased from 24 tons in 2008 to 4,700 tons in 2012. This was closely related to the rapid development of tetrachloroethylene production and the improvement of product quality in China.

5.5 1,1,1,2-Tetrafluoroethane

1,1,1,2-Tetrafluoroethane, also known as norfluran, R134a, 1,2,2,2-tetrafluoroethane and HFC-134a, is a derivative of trichloroethylene or tetrachloroethylene. CAS Registry Number: 811-97-2. Molecular formula: $C_2H_2F_4$, molecular weight: 102.0309. The chemical structure is

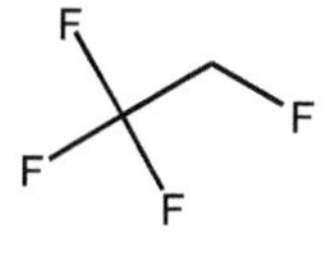

5.5.1 Properties of tetrafluoroethane

Tetrafluoroethane (HFC-134a) is a colorless and nonflammable chemical substance with a boiling point of −26.6 °C, a saturated vapor pressure of 661.9 kPa (25 °C) and an enthalpy of vaporization of 216 kJ kg^{-1} (1 atm). It is insoluble in water but soluble in ether. The lack of chlorine in the molecule reduces the ozone depletion activity to practically zero. The thermodynamic performance is very similar to CFC-12 (boiling point: −29.8 °C), and it is also comparable in safety to CFC-12. Therefore, it is regarded as the best alternative to CFC-12. Although HFC-134a has a certain greenhouse effect (HGWP = 0.28), which does not affect it as the preferred ODS (Ozone Depleting Substance) alternative. Since the mid-1990s, HFC-134a has been widely used in the industrial and commercial refrigeration, automotive air conditioning, central air conditioning, household refrigerators, plastic foam, pharmaceuticals, cosmetic aerosols and medical aerosol propellants. In recent years, tetrafluoroethane is also used as an extraction solvent.

5.5.2 Production process of tetrafluoroethane

Dozens of synthetic routes for tetrafluoroethane have been reported. Among them, comprehensive consideration of the sources of raw materials, production processes and waste treatments and other factors, only two technical routes, fluorination of trichloroethylene and fluorination of tetrachlorethylene, have been applied for industrial production of tetrafluoroethane. Trichloroethylene route is preferentially recommended because of its simple reaction step and less amounts of by-products [8, 9]. At present, trichloroethylene route has been used by tetrafluoroethane manufacturers such as ICI Company (UK), ATOFINA (France), Daikin (Japan), Osmont (Italy) and Hoechst (Germany). Only DuPont (United States) uses the tetrachloroethylene route for co-production of tetrafluoroethane with other products.

5.5.2.1 Trichloroethylene route

In the presence of a catalyst, fluorination of trichloroethylene with hydrogen fluoride generates 1,1,1-trifluoro-2-chloroethane (HCFC-133a); then, at higher temperature, fluorination of 1,1,1-trifluoro-2-chloroethane gives 1,1,1,2-tetrafluoroethane (HFC-134a):

$$\mathrm{ClCH{=}CCl_2 + 3HF \xrightarrow{Catalyst} ClCHCF_3 + 2HCl}$$

$$\mathrm{ClCHCF_3 + HF \xrightarrow{Catalyst} FCH_2CF_3 + HCl}$$

In the trichloroethylene route, there are three processes: liquid-phase process, liquid-phase–gas-phase process and gas-phase process.

5.5.2.1.1 Liquid-phase process

The advantage of the liquid-phase process is that it is following the traditional Freon production process with simple process and mature technology. DuPont Company developed liquid-phase fluorination for preparation of HFC-134a in 1982. However, due to the emergent corrosion of the equipment at high temperature and the difficulty in continuous operation, this method is still in the stage of laboratory test. Shanghai Institute of Organic Chemistry, a division from Chinese Academy of Science, improved the liquid-phase method by using perfluoroalkoxysulfonyl fluoride as the catalyst. The reaction was carried out in KF solution at 230 °C and 12.5 MPa for 2 h, resulting in a yield of 88% for HFC-134a. In the improved liquid-phase process, the reaction temperature is reduced, which is good for reduction of corrosion and byproducts, making continuous production possible, but it is still difficult to achieve industrial production in a short time.

5.5.2.1.2 Liquid-phase–gas-phase process

The disadvantage of the liquid-phase method is that the reaction is carried out in an autoclave, especially the second step of the equilibrium reaction, which requires a higher reaction temperature and pressure, resulting in great difficulties to equipment manufacturing and safe production; therefore, liquid-phase–gas-phase process was developed for the production of HFC-134a. In the liquid-phase–gas-phase process, the first step reaction is carried out in liquid phase to produce HCFC-133a, and the second step reaction is performed in gas phase to convert HCFC-133a into HFC-134a. The original equipment and techniques for manufacture of Freon products can be applied to the first step reaction, which is beneficial for those factories to change their product from CFCs to HFC-134a. However, in the second step reaction, the conversion is low and the catalyst life is short, which restricts this method to be applied for the large-scale production of HFC-134a.

5.5.2.1.3 Gas-phase method

Gas-phase method is that both the two step reactions are carried out in gas phase. It has advantages including easy reaction control, less amount of waste pollution and convenient large-scale continuous production. Gas-phase process has gradually replaced liquid-phase process and liquid-phase–gas-phase process and become the mainstream of HFC-134a production in the world. At present, many chemical companies in the world use this method to produce HFC-134a, such as ICI (United Kingdom), DuPont (United States) and Showa Denko (Japan).

In the gas-phase process, trichloroethylene reacts with anhydrous hydrogen fluoride in the presence of a chromium-based catalyst. HCFC-133a is obtained in the first step reaction by addition and substitution, and then it is further reacted with HF in the second step reaction to generate the final product HFC-134a at a temperature of 350–380 °C. The second step reaction is relatively difficult, and the conversion of trichloroethylene is generally only about 20%. Therefore, the continuous recycling of trichloroethylene is often used in industrial production to recover a large amount of raw materials, reduce toxic and harmful intermediates and increase the overall yield.

The process flow diagram for production of tetrafluoroethane by gas-phase process is shown in Figure 5.1.

Trichloroethylene reacts with hydrogen fluoride in reactor 1 to form HCFC-133a, and the reaction mixture enters reactor 2 for further reaction with hydrogen fluoride

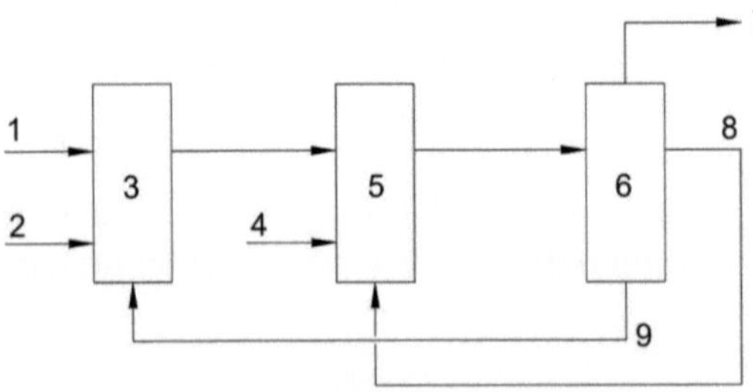

1. Trichloroethylene, 2, 4. Hydrogen fluoride, 3. Reactor 1, 5. Reactor 2, 6. Fractional distillation tower, 7. HFC-134a, 8. HFC-133a, 9. Bottom product of the tower.

Figure 5.1: Process flow diagram for production of tetrafluoroethane by gas-phase process.

to obtain HFC-134a. The reaction mixture is subjected to fractional distillation. The top product of the distillation tower is HFC-134a. The side fraction (8) mainly contains HCFC-133a, HFC-134a and a small amount of HF, which are sent back to the reactor 2 for reuse. The bottom product of the tower contains unreacted trichloroethylene and HF and returns to the reactor 1.

The synthesis of tetrafluoroethane from trichloroethylene is technically and economically suitable for China, especially the gas-phase method is easier to be industrialized.

5.5.2.2 Tetrachloroethylene route

Tetrafluoroethane is produced from tetrachloroethylene and hydrogen fluoride. The reaction equations are as follows:

$$Cl_2C{=}CCl_2 + Cl_2 + 3HF \longrightarrow Cl_2CFCClF_2 + 3HCl$$

$$Cl_2CFCClF_2 \longrightarrow CF_3CCl_3$$

$$CF_3CCl_3 + HF \longrightarrow CF_3CCl_2F + HCl$$

$$CF_3CCl_2F + H_2 \longrightarrow FCH_2CF_3 + Cl_2$$

Using chromium plus gold as the catalyst, the conversion of tetrachloroethylene per pass was 50% and the selectivity to tetrafluoroethane was 80%. DuPont developed three generations of synthetic technologies. Compared to the first-generation technology, the reactors and catalysts were improved and the production capacity of the reactor with the same volume was doubled in the second-generation technology. The third-generation technology does not require a catalyst, but requires strict temperature control.

This process produces both HFC-134a and HCFC-124. If HFC-134a is the only product required, all of the HCFC-124 are recycled. The moisture and nitrogen contents in the hydrogen should be less than 10×10^{-6} and $1{,}000 \times 10^{-6}$, respectively. The hydrogen is purified by pressure swing adsorption.

5.5.3 Uses of HFC-134a

HFC-134a is the most widely used medium- and low-temperature environmentally friendly refrigerant. Due to the good comprehensive performance of HFC-134a, it is a very effective and safe alternative to CFC-12 and mainly used as a refrigerant in refrigerators, freezers, air conditioners, dehumidifiers, cold storage, ice water machines, ice cream machines, refrigeration condensing units and so on. It can also be applied as aerosol propellant, polymer (plastic) physical foaming agent and magnesium alloy shielding gas.

5.5.4 Production status of HFC-134a

HFC-134a has been produced since the 1990s. The main manufacturers include DuPont and ICI, and the production output has been increased year by year. According to statistics, from 1995 to 2005, the average annual growth rate was 13%, and the output in 2005 was 170,000 tons.

In China, the production of HFC-134a started in 1999. At that year, Xi'an Institute of Modern Chemistry built a 200 t a^{-1} HFC-134a production line, ending the history of China's Freon substitutes relying entirely on imports from other countries. In December 2003, Xi'an Jinzhu Modern Chemical Co., Ltd. built a production line with an annual output of 5,000 t HFC-134a in Xi'an Weihe Industrial District, which was the largest production project for Freon substitute in the country at that time. In recent years, the production of HFC-134a has developed rapidly. At present, the main manufacturers include Xi'an Jinzhu Modern Chemical Co., Ltd., Quzhou Juhua Group, China Sinochem Group, Zhejiang Sanmei Chemical Co., Ltd. and Shandong Dongyue Chemical Co., Ltd., with a total production capacity of more than 100,000 t a^{-1}.

5.6 1,1-Difluoroethane

1,1-Difluoroethane, also known as R-152a, HFC-152a, is a derivative of acetylene and hydrogen fluoride. CAS Registry Number: 75-37-6. Molecular formula: $C_2H_4F_2$, molecular weight: 66.0500. The chemical structure is

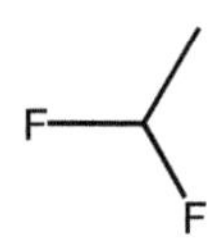

5.6.1 Properties of 1,1-difluoroethane

1,1-Difluoroethane is a colorless gas at room temperature. It can easily be liquefied. Its melting point and boiling point are –117 °C and –24.7 °C, respectively. The enthalpy of vaporization is 22.7 kJ mol^{-1}. It is insoluble in water. The explosion limit in air is from 3.7% to 18%.

5.6.2 Processes for manufacture of 1,1-difluoroethane

According to different raw materials, the processes for manufacture of 1,1-difluoroethane include two routes: fluorination of acetylene and fluorination of vinyl chloride.

5.6.2.1 Fluorination of acetylene

Acetylene undergoes a fluorination reaction with anhydrous hydrogen fluoride to produce 1,1-difluoroethane:

$$CH{\equiv}CH + 2HF \xrightarrow{\text{Catalyst}} CH_3CHF_2$$

The synthesis of 1,1-difluoroethane from acetylene and HF has the advantages including simple process, easily available raw materials and low cost. The reaction can be carried out either in liquid phase or in gas phase [10–12]. In the liquid phase, fluorosulfonic acid is used as the catalyst, the reaction is carried out at a temperature of 30 °C under a pressure of 0.03–0.3 MPa. In the gas phase, AlF_3 is generally used as a catalyst, and the reaction is carried out at 220–280 °C under atmospheric pressure. The gas-phase process has the advantages including higher catalytic activity and conversion (>98%). Moreover, the gas-phase process does not have problems such as high corrosion of the catalyst and difficulty in the treatment of the wastes as that in the liquid-phase process. However, the surface of the catalyst is easily coked during the reaction, resulting in catalyst deactivation. Addition of Mn element to the catalyst improved the reaction selectivity slightly. The coking of the catalyst can be suppressed to some extent by addition of Bi element to the catalyst.

In China, liquid-phase process has been used to produce 1,1-difluoroethane for more than 30 years, and the process is quite mature. The gas-phase method appears late, and the process needs to be improved and matured. However, from the perspective of equipment corrosion and environmental protection, the gas-phase process has advantages and deserves further development.

5.6.2.2 Fluorination of vinyl chloride

The raw materials for production of 1,1-difluoroethane by the fluorination of vinyl chloride are vinyl chloride and anhydrous hydrogen fluoride. The reaction process includes addition and substitution. Vinyl chloride undergoes an addition reaction with

HF to generate CH_3CHClF, and then the resultant CH_3CHClF reacts with HF, leading to fluorine atom in substitution for chlorine atom and generation of 1,1-difluoroethane:

$$CH_2{=}CHCl + HF \xrightarrow{\text{Cat.}} CH_3CHClF$$

$$CH_3CHClF + HF \xrightarrow{\text{Cat.}} CH_3CHF_2 + HCl$$

The catalyst is generally selected from the halides of titanium, vanadium, tantalum, molybdenum, tungsten, tin and stibium. It is recommended to use $SnCl_4$ as a catalyst. During the operation, hydrogen fluoride and vinyl chloride are continuously introduced into the reaction vessel, and the reaction mixture is continuously discharge from the reactor. The residence time is generally 0.1–5 h. Although industrial production of 1,1-difluoroethane by fluorination of vinyl chloride has accounted for a large proportion, the process still has the following problems: First, the service life of the catalyst is short, the production capacity is small, and the unit consumption is high. It is difficult to increase the capacity of a single kettle. Second, severe fouling of carbonaceous solids on the reactor wall occurs because of a large amount of high boiling by-products such as polymers and oily substances. Third, the conversion of vinyl chloride is not high enough, and the unreacted vinyl chloride forms an azeotrope with 1,1-difluoroethane, which makes it difficult to purify 1,1-difluoroethane and thus limits the application of 1,1-difluoroethane in many fields.

Some efforts have been made to solve the above problems. Huang et al. [13] optimized and improved the fluorination process, so that the efficiency of the catalyst (reaction material mass/catalyst mass) increased from 60–80 g g^{-1} to 130–150 g g^{-1}; vinyl chloride feed rate from 0.1–0.18 h^{-1} was increased to 0.28–0.32 h^{-1}, the productivity of a single kettle was greatly improved, and the carbonaceous solid matter on the reactor wall and the high-boiling substances such as polymers and oily substances formed by the reaction were greatly reduced. The product quality has been significantly improved. Chen [14] proposed a two-step method to produce 1,1-difluoroethane, which improved the conversion and selectivity of the reaction. The conversion of vinyl chloride was complete, so that there is no need to separate or use photochlorination and other methods to remove vinyl chloride.

5.6.3 Uses of 1,1-difluoroethane

5.6.3.1 For refrigerant

The production and use of chlorofluorocarbons have been limited because of the Montreal Protocol. HFC-152a has an ozone depletion potential (ODP) of zero, which is receiving increasing attention as a substitute for chlorofluorocarbons. For example, on 12 June 2008, the US Environmental Protection Agency announced that HFC-152a was included in the list of alternative refrigerants for automotive air conditioning.

HFC-152a can be used alone as a refrigerant or mixed with other refrigerants. For example, a mixture of HFC-152a and F-12 in a mass ratio of 26.2:73.8 has greater refrigeration capacity. The refrigerant mixture of HFC-152a and HFC-134a has better environmental performance and thermodynamical performance than HFC-134a alone, which can reduce energy consumption and cost.

5.6.3.2 For foaming agent

In view of the need for environmental protection, chlorofluorocarbons, which have traditionally been used for foaming agents, will be eliminated. Hydrocarbon foaming agents are specified as photochemically reducing volatile organic compounds (VOCs), and their use is also severely restricted. Therefore, 1,1-difluoroethane is often used as a foaming agent in place of trichlorofluoromethane for foaming of plastic products.

5.6.3.3 For the manufacture of other products

1,1-Difluoroethane is also used as an intermediate in organic synthesis for the synthesis of vinylidene fluoride, an important monomer for fluororubbers and fluoroplastics. For example, 1,1-difluoroethane is used for production of PVDF resin, and typical manufacturers include Pennwalt Company (United States). In China, Shanghai Institute of Organic Materials, Chenguang Chemical Research Institute and Zhejiang Research Institute of Chemical Industry are also producing PVDF resin from 1,1-difluoroethane. The production output of PVDF resin occupies the second place in fluororesin. With the development and application of PVDF resin in recent years, the market demand for 1,1-difluoroethane is increasing year by year in China.

5.6.3.4 For heat pump working fluid

1,1-Difluoroethane can be used as a working fluid for heat pump. The two materials, HFC-152a and 1,1,1,3,3,3-hexafluoropropane, are physically mixed at a room temperature to form a corresponding mixed working medium, which is suitable for medium and high temperature heat pump units. The mixed working medium does not destroy the ozone layer, meeting environmental protection requirements. It has suitable thermal parameters and excellent cycle performance [15].

References

[1] Zhang X. Production and application of tetrabromoethane in China, Liming Chemical 1996, 5, 12–13.
[2] Li P, Zheng Y, Chen M, et al. Effect of stabilizer on stability of tetrabromoethane, Contemporary Chemical Engineering 2011, 40(1), 30–32.
[3] Liu L. Study on the synthesis of tetrabromoethane, Shandong Chemical Industry 1999, 2, 3–4.
[4] Liu L, Li F. Study on the improvement of tetrabromoethane synthesis process, Haihu Salt and Chemical Industry 2003, 32(3), 21–22.

[5] Ding P, Wang X. Production technology and market analysis of trichloroethylene and tetrachloroethylene, Chlor-Alkali Industry 2012, 12, 1–5.
[6] Yao L. Production, consumption and market prospect of trichloroethylene, Guangzhou Chemical 2013, 7, 42–44.
[7] Miao N, Fu B, Wang J, et al. A method for stabilize tetrachloroethylene, Chlor-Alkali Industry 2014, 9, 34–38.
[8] Chen K, Ma X. Development of 1,1,1,2-tetrafluoroethane, Organofluorine Industry 2005, 3, 16–20.
[9] Dong H, Yu S. Synthesis of 1,1,1,2-tetrafluoroethane with composite catalyst, Organofluorine Industry 2007, 3, 13–15.
[10] Yu Q, Huang X. Study on vapor phase catalytic fluorination of acetylene to difluoroethane, Shanghai Chemical Industry 1993, 1, 12–16.
[11] Su L, He A. Research progress in production technology of difluoroethane (HFC-152a), Organic Fluorine Industry 2011, 4, 30–33.
[12] Liu J, Wang J, Yuan J. Production and application of 1,1-difluoroethane, Chemical Production and Technology 2011, 2, 6–8.
[13] Huang Y, Chen J. Optimization of fluorination process for production of HFC-152a by ethylene chloride method, Zhejiang Chemical 2008, 39(10), 8–9.
[14] Chen G. Study on the two-step synthesis of HFC-152a from vinyl chloride, Zhejiang Chemical Industry 2007, 38(9), 7–9.
[15] Wang H, Ma L, Zheng C. Medium and high temperature heat pump mixed working medium containing difluoroethane (HFC-152a). CN 200410077955.3, 2008.

Chapter 6
Derivatives from acetylene reacting with acetic acid

6.1 Vinyl acetate

Vinyl acetate, also known as acetic acid vinyl ester, ethenyl acetate, ethenyl ethanoate, vinyl acetate monomer and acetoxyethene, is a derivative of acetylene and acetic acid. CAS Registry Number: 108-05-4. Molecular formula: $C_4H_6O_2$, molecular weight: 86.0892. The chemical structure is

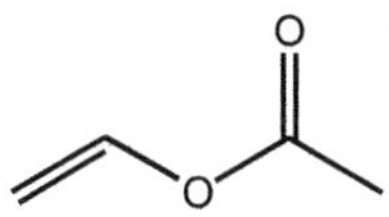

6.1.1 Properties of vinyl acetate

Vinyl acetate is a colorless, transparent and flammable liquid with a strong odor. It is soluble in most organic solvents and slightly soluble in water. The physical properties of vinyl acetate are listed in Table 6.1.

Vinyl acetate is anesthetic and irritating, causing inflammation of the mouth and red spots on the eyes in the air with high concentrations of its vapor. The toxicity of this product is low, and the oral LD_{50} of rats is 2,920 mg kg^{-1}.

Table 6.1: Physical properties of vinyl acetate.

Melting point/°C	−100.2	Enthalpy of vaporization (309 K)/ kJ · mol^{-1}	34.4
Boiling point/°C	72.5	Enthalpy of fusion (180.6 K)/kJ · mol^{-1}	8.46
Density (25 °C)/g · cm^{-3}	0.934	Heat capacity (25 °C)/J · mol^{-1} · K^{-1}	169.5
Refractive index (20 °C)	1.3950	Critical temperature/°C	245.98
pH (20 g/L, H_2O, 20 °C)	7	Critical pressure/MPa	4.185
Solubility in water (20 °C)/g · L^{-1}	23		

6.1.2 Process for manufacture of vinyl acetate

The industrial processes for production of vinyl acetate include acetylene route and ethylene route. Currently, ethylene route accounts for more than 70% of the total production capacity of vinyl acetate in the world. However, in areas with abundant coal and natural gas resources, the acetylene route to vinyl acetate is quite competitive. In China, due to the abundant coal resources and relatively insufficient petroleum

https://doi.org/10.1515/9783110714999-006

resources, acetylene route dominates the production of vinyl acetate, accounting for about 80% of the total production capacity [1–7].

6.1.2.1 Acetylene route

Acetylene undergoes an addition reaction with acetic acid vapor in the presence of a zinc acetate catalyst supported on activated carbon to produce vinyl acetate. The reaction is carried out at a temperature of 170–230 °C under a pressure of 35–40 kPa. The reaction equation is as follows:

$$HC{\equiv}CH + CH_3COOH \longrightarrow CH_3COOCH{=}CH_2$$

The process flow diagram of vinyl acetate production is shown in Figure 6.1. Acetic acid is continuously fed to an acetic acid evaporator where it is contacted with purified fresh anhydrous acetylene and recycled acetylene. The operating temperature and pressure of the evaporator are 70–80 °C and 27–34 kPa, respectively. The mixture of acetic acid vapor and acetylene (molar ratio 1:4–5) is heat exchanged with the hot gas from the reactor, and subsequently heated to 170–195 °C by steam, and then introduced into the vinyl acetate synthesis reactor filled with zinc acetate catalyst supported on activated carbon. In the reactor, acetylene reacts with acetic acid to form vinyl acetate. The outlet gas mixture from the reactor, which mainly contains vinyl acetate, unreacted acetic acid, acetylene and by-products, is partially condensed by heat exchange with the feed mixture, and further condensed by a condenser. Then, the resultant condensate is introduced into a gas–liquid separator for phase separation. The gas phase from the top of the gas–liquid separator is sent to a washing tower to remove and recover the small amount of acetic acid and vinyl acetate in it by washing with recycled acetic acid, and then recycled to the evaporator by a compressor. The liquid phase from the bottom of the gas–liquid separator and the liquid from the bottom of the washing tower are sent to the light fraction removal tower to remove the light component (acetaldehyde). The bottom product of the light fraction removal tower is introduced into the vinyl acetate fractional distillation tower. Vinyl acetate with a purity of 99.9% is obtained from the top product of the tower. The bottom product of the vinyl acetate fractional distillation tower is sent to acetic acid purification tower to remove crotonaldehyde from the top of the tower. The bottom product of the acetic acid purification tower is introduced into the heavy fraction removal tower to remove the heavy component, and the acetic acid with a purity of more than 99% is obtained at the top of the tower, which is recycled to the washing tower and the evaporator.

In order to avoid the polymerization of vinyl acetate, hydroquinone was added as a polymerization inhibitor to the feeds of the light fraction removal tower and the vinyl acetate fractional distillation tower.

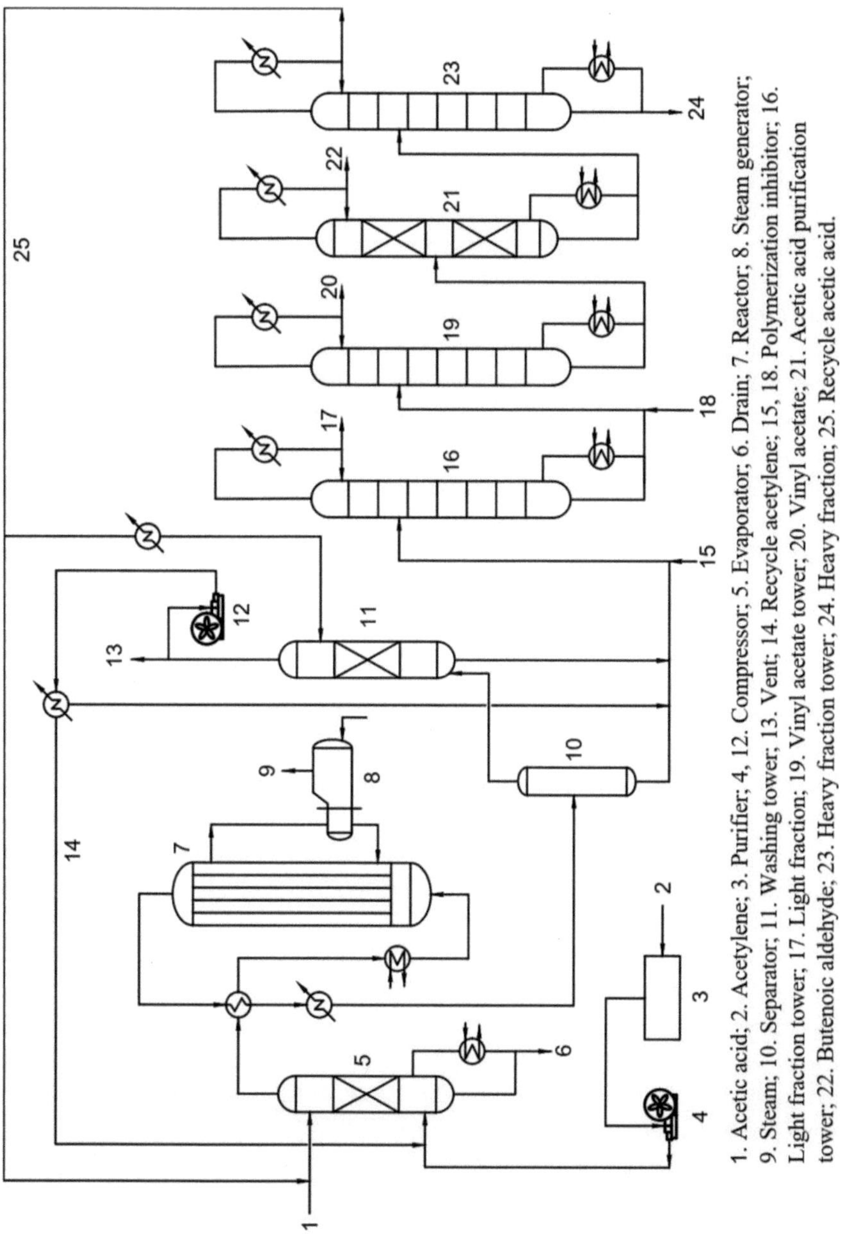

Figure 6.1: Process flow diagram of vinyl acetate production.

The main by-products are acetaldehyde and crotonaldehyde. The conversion of acetylene per pass is in the range from 60% to 75%, the selectivity to vinyl acetate is 99% based on acetic acid, and 93% based on acetylene.

The reactor for synthesis of vinyl acetate can be a fixed bed or a fluidized bed. The advantage of the fluidized bed reactor is that the temperature distribution in the reactor is uniform, and the temperature is reduced by 10 to 20 °C in comparison with the fixed-bed reactor, resulting in less acetylene polymer generation, longer catalyst life and better quality of vinyl acetate. The disadvantage of the fluidized bed reactor is that the wear of the catalyst is serious, and the wear-resistant activated carbon is required as a carrier.

The rapid development of petrochemical industry in the 1970s caused most of the vinyl acetate production that used the acetylene route to be replaced by the ethylene route because ethylene was cheaper than acetylene from calcium carbide. However, in recent years, the production of vinyl acetate by acetylene route has regained its vitality in the economy. In areas with abundant natural gas resources, acetylene from natural gas is cheaper than the acetylene from calcium carbide, leading to better economic benefits for the production of vinyl acetate. Currently, acetylene route is the main method for production of vinyl acetate in China.

6.1.2.2 Ethylene route

Ethylene reacts with oxygen and acetic acid in the presence of Pd–Au (or Pt) catalyst at a temperature of 100–200 °C under a pressure of 0.6–0.8 MPa, and the resultant reaction mixture is subjected to separation for vinyl acetate product.

The ethylene route can be performed in liquid phase or in gas phase. However, the catalyst system used in liquid phase contains chloride ions, which is highly corrosive to equipment, resulting in a short operation time. Therefore, liquid phase process has been eliminated.

The traditional ethylene route mainly includes Bayer process and USI process, both of which adopt a gas phase fixed-bed process, and the technical processes are also similar. The difference between them is that the catalyst used in Bayer process is Pd–Pt supported on silica gel, while the catalyst used in USI process is Pd–Au supported on α-Al_2O_3. Bayer process is more popular in industrial production because of higher conversion per pass of acetic acid and better selectivity to vinyl acetate than those of USI process.

6.1.2.3 Technical progress

Although the gas phase process of ethylene to produce vinyl acetate is technically improved compared with the acetylene process, the technology still has many defects in terms of reaction kinetics and catalyst deactivation mechanism, such as continuous deactivation of the catalyst and uneven distribution of reactor bed layer. The conversion of ethylene per pass is limited. In response to the above shortcomings, BP Amoco and Celanese have improved the traditional process and successfully developed a new production process, respectively.

BP Amoco has developed the Leap process for the production of vinyl acetate from ethylene by the fluidized bed gas-phase process. The process uses a newly designed fluidized bed reactor system and catalyst, which reduces the investment cost of the device by 30%. The reaction is carried out at 152 °C and 0.9 MPa, and the conversions of ethylene, oxygen and acetic acid per pass are 8.3%, 30.3% and 41.25%, respectively.

The fixed-bed gas phase process developed by Celanese Company is also called the Vantage process. It still uses a fixed-bed reactor. However, due to important improvements in the catalyst system and a new technology for the recovery of ethylene in the tail gas, it overcomes the defects of the traditional fixed-bed process, so the yield of vinyl acetate is significantly enhanced. The Vantage process also uses a Pd-Au catalyst system, but uses sodium citrate as a reducing agent and silica gel as a carrier to prepare metal catalyst with a particle size of 20 nm. Furthermore, lanthanide rare earth elements such as praseodymium and neodymium are added to PdAu/K catalyst, thereby significantly improving the activity and selectivity of the catalyst and the space–time yield. Using this technology to expand the capacity of the original fixed-bed reactor, the investment is only equivalent to 10–15% of the investment for a new device of the same scale, and the production cost can be reduced by 2–8%. Therefore, the Vantage process developed by Celanese Company occupies a leading position in the fixed-bed ethylene gas phase process.

In China, Sichuan Vinylon Factory has improved the technology of acetylene process, upgrading the equipment capacity from 90,000 to 210,000 t a^{-1}. Its process technology of vinyl acetate production from natural gas acetylene is at the world advanced level. Some technical progress has also made in ethylene process. Shanghai Research Institute of Petroleum and Chemical Industry developed CTV series catalyst. The activity of CTV-IV catalyst exceeds that of Bayer-II catalyst by 13.8% while the selectivity to vinyl acetate maintains at 91.7%, showing an overall performance superior to Bayer-II catalyst. In the industrial application, vinyl acetate output of the device is increased by 8.75% after using CTV-IV catalyst in place of Bayer-II catalyst.

6.1.3 Uses of vinyl acetate

Vinyl acetate is mainly used for production of polyvinyl acetate (PVAc), polyvinyl alcohol (PVA), ethylene–vinyl acetate (VAE) copolymer, vinyl chloride–vinyl acetate copolymer and other derivatives. These derivatives are used in chemical products such as coatings, adhesives, vinylon, films and vinyl copolymer resins.

In China, the consumption of vinyl acetate in PVA production is the largest, accounting for about 65% of the total amount of vinyl acetate. PVAc is the second largest consumer for vinyl acetate, accounting for about 16% of total consumption. VAE consumes about 11% of vinyl acetate. Other consumption accounts for about 8%.

6.1.4 Production status of vinyl acetate

The global production output of vinyl acetate is about 6 million tons per year. Currently, Celanese Corporation is the largest manufacturer of vinyl acetate, and its production capacity accounts for approximately 23.14% of the total production capacity of vinyl acetate in the world. The next is Dalian Chemical Company (Taiwan, China), whose production capacity accounts for about 9.49% of the global production capacity. The third largest manufacturer of vinyl acetate is Sinopec Group (including Sinopec Sichuan Vinylon Factory, Sinopec Shanghai Petrochemical Company, and Sinopec Beijing Dongfang Petrochemical Company), accounting for approximately 6.39% of the total production capacity in the world.

In China, 14 manufacturers are engaging in the production of vinyl acetate, with a total output of more than 1.2 million tons/year. There are 12 production enterprises using acetylene route, of which 11 are using acetylene from calcium carbide as raw material, and only Sinopec Sichuan Vinylon Factory is using acetylene from natural gas as raw material. Sinopec Shanghai Petrochemical Company and Sinopec Beijing Dongfang Petrochemical Company use the ethylene route for the production of vinyl acetate.

6.2 Polyvinyl acetate

PVAc, also known as poly(vinyl acetate), is a homopolymer of vinyl acetate. CAS Registry Number: 9003-20-7. The chemical structure is

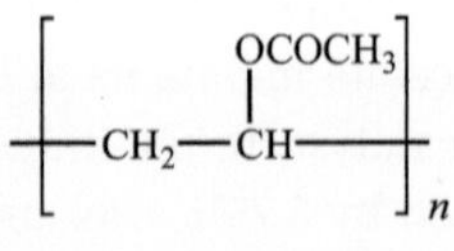

6.2.1 Properties of polyvinyl acetate

PVAc is a transparent glassy substance at room temperature. It is a colorless, odorless, nontoxic, thermoplastic and water-insoluble polymer. In the presence of an acid or alkali, it can be alcoholyzed to produce PVA.

PVAc has high stability to light and hardly ages under the action of light. If it is heated to above 150–190 °C, it will start to decompose into acetic acid and nonvolatile hydrocarbons.

The glass transition temperature T_c of PVAc varies with the molecular weight. For example, $T_c = 17$ °C at the molecular weight of 15,000, and $T_c = 26$ °C at the molecular weight of 104,000.

PVAc is insoluble in water and only slightly swells in water.

PVAc emulsion is commonly known as white latex. It can be directly modified with a variety of additives. It has good mechanical strength and no lack of glue. In particular, it is a water-based adhesive and has no environmental pollution.

6.2.2 Production process of polyvinyl acetate

PVAc is a high molecular compound formed by free radical polymerization of vinyl acetate. The polymerization reaction is as follows:

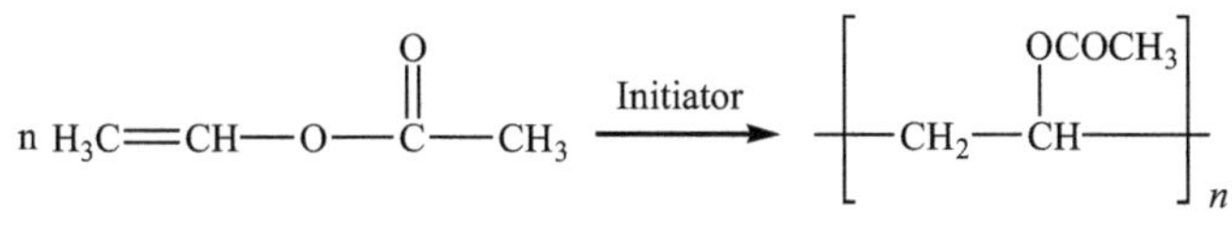

The polymerization processes include bulk polymerization, solution polymerization, suspension polymerization, and emulsion polymerization. Usually, solution polymerization and emulsion polymerization are used for production of PVAc [8, 9]. For example, using dibenzoyl peroxide as an initiator and methyl ethyl ketone as a solvent, PVAc is prepared by solution polymerization at a temperature of 65 °C for about 10 h.

6.2.3 Uses of polyvinyl acetate

The major uses of PVAc are in water-based latex paints, adhesives, fabric finishes, and lacquers. PVAc is also used as base material for gum-based candy in food industry [9].

6.2.3.1 Homopolymerized polyvinyl acetate emulsion adhesive

Adhesion by means of PVAc emulsion has important practical significance. The reason is that this emulsion has special adhesion and viscosity, is nontoxic, nonflammable and can be diluted with water, and has good re-emulsification properties, and the most prominent is its fast-curing characteristics. The property of fast curing is very important for accelerating the bonding process and adopting the corresponding fast coater.

PVAc emulsion adhesives are used in many industrial sectors, such as in the wood processing industry (plywood and furniture production), packaging, printing (binding) and construction.

PVAc emulsion can directly be spray dried to a loose powder adhesive. The main advantages of powder adhesive are convenient for transportation and storage. The powder can be mixed with water at room temperature to form an emulsion. The performance of the emulsion prepared by the powder is almost the same as that of the original emulsion.

6.2.3.2 Vinyl acetate copolymer adhesive (emulsion glue)

Among vinyl acetate copolymers, the most widely used adhesive is VAE copolymer. The difference between this copolymer and homopolymerized PVAc emulsion is that it can form a sufficiently soft film and does not require a plasticizer.

Compared with other monomers, ethylene has many advantages as a plasticizable comonomer. First of all, it is one of the cheapest products. Next, VAE copolymer emulsion adhesive requires a lower temperature to form a film than homopolymer emulsion adhesive.

The characteristics of emulsion glue are fast curing, good stability, easy to use and especially wide application range. The adhesion performance of VAE emulsion to many surfaces is stronger than that of vinyl acetate and alkyl maleate, alkyl fumarate and alkyl acrylate copolymers. It has good adhesion even without a tackifier. This adhesive is resistant to alkali, acid, oxygen, ozone and ultraviolet radiation. It is physiologically harmless to humans and is relatively inexpensive. It can bind various materials, including non-coated paper, cardboard and plastic films.

6.3 Polyvinyl alcohol

PVA is a water-soluble thermoplastic polymer prepared by partial or complete hydrolysis of PVAc with methanol or water. CAS Registry Number: 9002-89-5. The chemical structure is

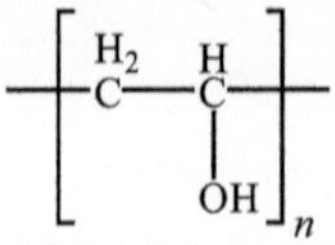

6.3.1 Properties of polyvinyl alcohol

PVA is an odorless white flake, flocculent or powdery solid. It is soluble in water, slightly soluble in dimethyl sulfoxide, but insoluble in gasoline, kerosene, vegetable oil, benzene, toluene, dichloroethane, carbon tetrachloride, acetone, ethyl acetate, methanol, ethylene glycol and so on.

The degree of polymerization of PVA is divided into ultra-high degree of polymerization (molecular weight of 250,000–300,000), high degree of polymerization (molecular weight of 170,000–220,000), medium degree of polymerization (molecular weight of 120,000–150,000) and low degree of polymerization (25,000–35,000). The degrees of alcoholysis are generally 78%, 88% and 98%. The degrees of alcoholysis for partial alcoholysis are usually 87–89%, and the degree of alcoholysis for complete alcoholysis is 98–100%. For product type of PVA, the thousand and hundred digits of the average degree of polymerization are usually put in the front, and the percentage of alcoholysis degree is put in the back. For example, PVA 17-88 means PVA with a polymerization degree of 1,700 and an alcoholysis degree of 88%. In general, as the degree of polymerization increases, the viscosity of the aqueous solution increases, and the strength and solvent resistance after film formation increase, but the solubility in water and the elongation after film formation decrease.

The relative density of PVA (25 °C/4 °C) is 1.27–1.31 (solid) and 1.02 (10% solution). It has a melting point of 230 °C and a glass transition temperature of 75–85 °C. It will change color when heated to above 100 °C in the air. When heated to 160–170 °C, dehydration and etherification occurs and the solubility is lost. PVA starts to decompose at 200 °C.

6.3.2 Process for manufacture of polyvinyl alcohol

PVA prepared by partial or complete hydrolysis of PVAc with methanol [10]. In the presence of sodium hydroxide catalyst, three reactions in the methanol solution of PVAc occur as follows:

Transesterification reaction:

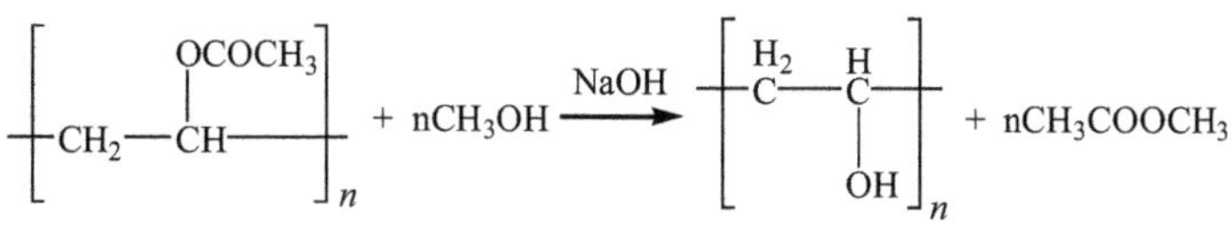

Saponification reaction:

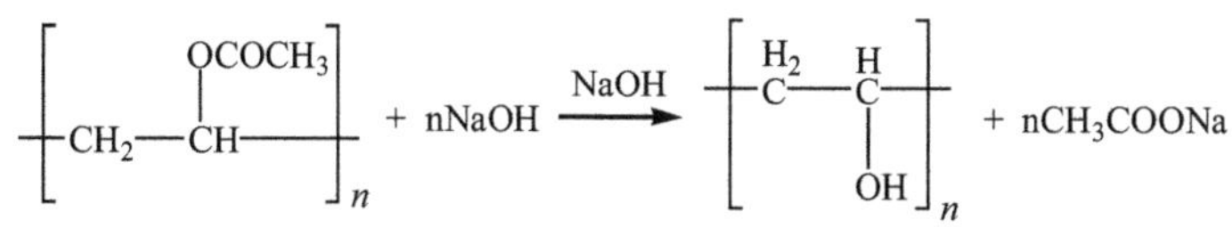

Side reaction:

$$CH_3COOCH_3 + NaOH \longrightarrow CH_3OH + CH_3COONa$$

In the production process of PVA, there are two kinds of alkaline alcoholysis processes, wet process and dry process, which are often referred to as high alkali method and low alkali method. In wet alcoholysis process, the methanol solution of PVAc contains 1–2% of water, and the sodium hydroxide catalyst is prepared in the form of an aqueous solution. The amount of the alkali used is large and the reaction solution has higher alkali concentration. Therefore, it is also known as high alkali method. The advantage of high alkali method is that the speed of alcoholysis is faster and the production capacity of the equipment is higher. However, the disadvantage is that side reaction for generation of sodium acetate increases. The content of sodium acetate in PVA product is high, resulting in low purity of PVA and high ash content, which affects the quality of the product.

Dry alcoholysis means that the water content in the methanol solution of PVAc is less than 1%, and the alcoholysis is carried out almost in the absence of water. The amount of alkali used is small, and the consumption of sodium hydroxide is only 1/10 of that in the wet alcoholysis process. Moreover, the side reaction to sodium acetate is greatly decreased. Therefore, dry alcoholysis has become the main process for production of PVA. However, alcoholysis reaction is slower in the dry alcoholysis process.

In the dry alcoholysis process, the concentration of PVAc in the methanol solution is about 35%, and the reaction is carried out at a temperature of 40 °C for 15–20 min.

6.3.3 Uses of polyvinyl alcohol

PVA is used in the manufacture of construction glue, polyvinyl acetal, vinylon, fabric treatment agent, dispersant, paper coating, adhesive and so on.

6.3.4 Production status of polyvinyl alcohol

In China, the total production capacity of PVA is 1.371 million t a^{-1}, including the capacities of some closed factories because of small scale and high cost. The production capacity using calcium carbide acetylene method is 960,000 t $year^{-1}$, accounting for 70.02% of the total production capacity; the production capacity using petroleum ethylene method is 251,000 t a^{-1}, accounting for approximately 18.31% of the total production capacity; and the production capacity using natural gas acetylene method is 160,000 t a^{-1}, accounting for approximately 11.67% of the total production capacity. Therefore, more than 80% of PVA capacity is from acetylene. Anhui Wanwei High-Tech Materials Co., Ltd. is currently the largest producer of PVA, with a production capacity of 350,000 t a^{-1} (including Inner Mongolia Mengwei and Guangxi Guangwei, which now belong to Anhui Wanwei) in 2018, accounting for approximately 25.53% of total production capacity; next is the Sichuan Vinylon Plant of Sinopec, with a production capacity of 160,000 t a^{-1}, accounting for approximately 11.67% of the total production capacity. Table 6.2 lists the main manufacturers of PVA in China.

The apparent consumption of PVA was 481,600 tons in 2005, 538,400 tons in 2010, 650,800 tons in 2015 and 639,400 tons in 2018. The demand for PVA for polymerization additives accounts for about 37.0% of the total consumption, fabric slurry accounts for about 20.0%, architectural coatings account for about 6.0%, vinylon fiber accounts for about 13.0.%, paper slurry and coating account for about 8.0%, adhesives account for about 11.0% and other aspects account for about 5.0% [11].

Table 6.2: The main manufacturers of polyvinyl alcohol in China.

No.	Manufacturer	Capacity/(1,000 t a^{-1})	Production process
1	Inner Mongolia Mengwei Technology Co., Ltd.	200	Calcium carbide acetylene route
2	Sinopec Sichuan Vinylon Plant	160	Natural gas acetylene route
3	Inner Mongolia Shuangxin Environmental Protection Material Co., Ltd.	130	Calcium carbide acetylene route
4	Ningxia Dadi Chemical Co., Ltd.	130	Calcium carbide acetylene route
5	Changchun (Jiangsu) Chemical Co., Ltd.	120	Petroleum ethylene route
6	Anhui Wanwei High-Tech Materials Co., Ltd.	100	Calcium carbide acetylene route
7	Shanxi Sanwei Group Co., Ltd.	100	Calcium carbide acetylene route
8	Sinopec Great Wall Energy Chemical (Ningxia) Co., Ltd.	100	Calcium carbide acetylene route
9	Guangxi Wanwei Biomass Technology Co., Ltd.	50	Bio-ethylene route
10	Sinopec Shanghai Petrochemical Co., Ltd.	46	Petroleum ethylene route
11	Beijing Dongfang Petrochemical Co., Ltd.	35	Petroleum ethylene route

6.4 Polyvinyl butyral

Polyvinyl butyral (PVB) is a synthetic resin obtained by the condensation reaction of PVA and butyraldehyde in the presence of acid catalyst. Since the acetalization reaction cannot be carried out completely, the molecular chain of PVB contains three functional groups, butyral, alcohol hydroxyl and acetate, the latter two of which come from PVA. CAS Registry Number: 63148-65-2. The chemical structure is

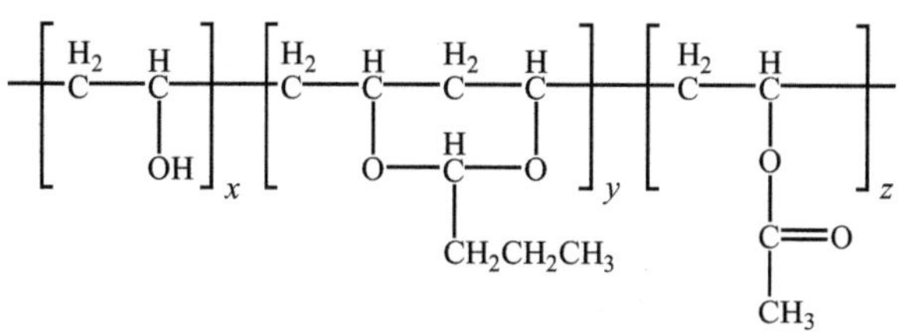

The proportion of these three groups can be controlled by changing the reaction conditions, but they are randomly distributed in the long molecular chain. The proportion and distribution of the three are different, and the properties of PVB are also different.

6.4.1 Properties of PVB

PVB resin is white or slightly yellow amorphous powder, with a particle size of 40–200 μm, fluidity and a density of 1.11 kg m^{-1}. It is a thermoplastic material, the softening temperature is 60–65 °C, the glass transition temperature is 70–78 °C, the molding temperature is 120–170 °C, and the temperature adaptation range is 10–80 °C

PVB resin is insoluble in water, but soluble in a variety of organic solvents such as alcohols, esters, benzene and ketones. It is miscible with various plasticizer including phthalates, phosphates, and fatty acid ester. It has good compatibility with various resins such as phenolic resin, urea–formaldehyde resin, melamine resin, epoxy resin, nitrocellulose and natural resin. By mixing these resins in an appropriate ratio, the physical and chemical properties are improved, and the cost is reduced.

PVB resin has certain chemical stability, but it can be decomposed in acidic or alkaline (especially acidic) medium. It starts to decompose thermally after being heated to 100 °C, and will be almost completely decomposed at 200–240 °C. Hydrolysis can occur in dilute acid, at this time intermolecular cross-linking can occur, a small amount of double bonds is formed in the chain, and the acetate group will be further hydrolyzed. Heated in concentrated acid to form a transparent solution, while decomposing into aldehydes. Upon heating in dilute alkali solution with alkali concentration less than 1%, the free acid in PVB resin is neutralized and the thermal stability of the resin is improved. However, when PVB resin is heated with 3% alkaline solution to boiling, the acetate groups can be further hydrolyzed and saponified.

6.4.2 Process for manufacture of PVB

PVB is obtained by the aldol condensation reaction of PVA and *n*-butyraldehyde in the presence of acid catalyst:

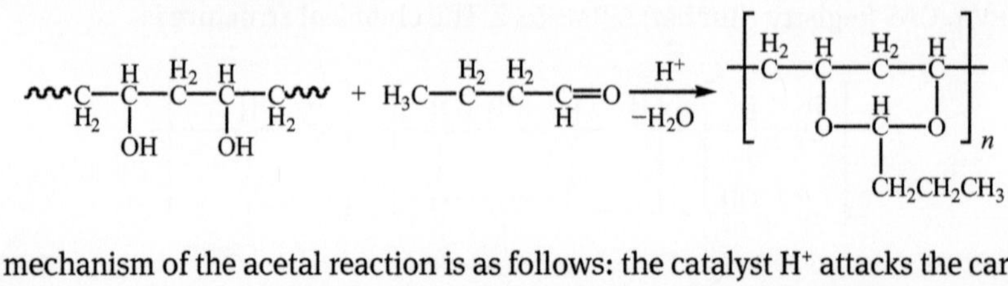

The mechanism of the acetal reaction is as follows: the catalyst H^+ attacks the carbonyl carbon in the *n*-butanal molecule to generate an active group with positive charge. The resultant active group attacks a hydroxyl in the PVA molecule, and one molecule of water is removed. Subsequently, the active group is combined with another adjacent hydroxyl to generate a six-membered ring acetal structure, and hydrogen ion is released.

Currently, industrial production processes of PVB include one-step method and two-step method. Different starting materials are used in these methods. PVAc is used as the starting material in one-step method, while PVA is used as the starting material in two-step method [12, 13].

6.4.2.1 One-step method

This is a production process in which the solvolysis of PVAc and the acetalization reaction with *n*-butyl aldehyde are carried out simultaneously to form PVB. Catalyst that accelerates solvolysis also promotes the acetalization reaction. The reaction can be carried out in a hydrophilic solution or in a hydrophobic solution. Considering the phase state of the solution in the later stage of the reaction, it can be divided into dissolution method and precipitation method.

6.4.2.1.1 One-step dissolution method

The one-step dissolution method is represented by the production process of Monsanto Company, and its process flow diagram is shown in Figure 6.2. Vinyl acetate undergoes a bulk polymerization to form a methanol solution of PVAc, and hydrochloric acid is added to the solution for alcoholysis of PVAc. The resultant PVA is suspended in the solvent. *n*-Butyraldehyde and hydrochloric acid are added to the suspension and acetalization reaction occurs. The reaction is carried out at reflux temperature for 8–10 h. As the reaction proceeds, the PVA is gradually consumed and finally the reaction mixture forms a homogeneous solution. Then, the pH value of the solution is adjusted to be 6 by addition of alkali. Water is added to the solution, resulting in PVB precipitate. The precipitate is filtrated, washed and dried to obtain the PVB finished product.

6.4.2.1.2 One-step precipitation method

Vinyl acetate is bulk polymerized to produce PVAc, which is mixed with hydrochloric acid and hydrolyzed. When the hydrolysis is nearly finished, *n*-butyl aldehyde is added to the reaction solution, and the acetalization reaction is rapidly carried out with vigorous stirring in the presence of acid catalyst. The PVB formed is insoluble and precipitated from the reaction solution. After neutralization of the reaction mixture, the precipitate is filtrated, washed and dried to obtain the PVB finished product. Because of the high speed of the acetalization reaction, the PVB precipitated as a dense hard block, and the impurities in the PVB product were difficult to remove by washing. Therefore, the process is rarely used in industry.

6.4.2.2 Two-step method

PVA used as the starting material, and PVB resin was prepared by reacting PVA solution with *n*-butyraldehyde in the presence of a suitable acid catalyst. According to the different solvents used, the reaction can be carried out in homogeneous system or in

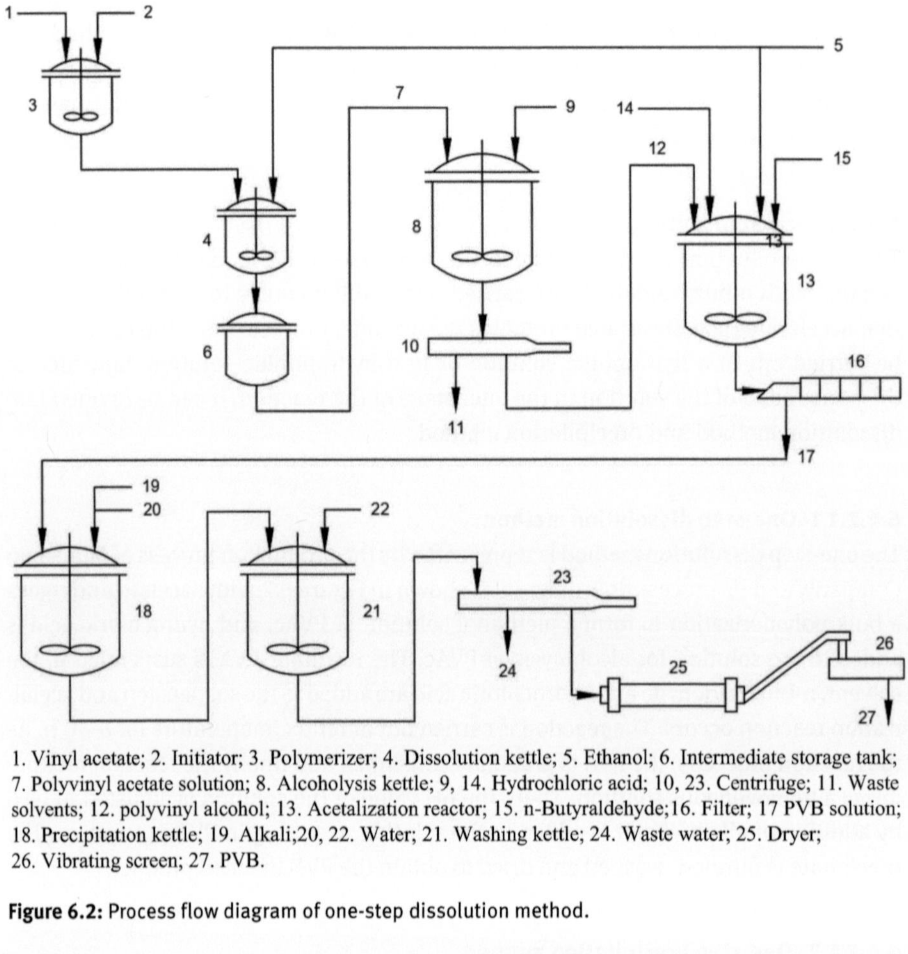

1. Vinyl acetate; 2. Initiator; 3. Polymerizer; 4. Dissolution kettle; 5. Ethanol; 6. Intermediate storage tank; 7. Polyvinyl acetate solution; 8. Alcoholysis kettle; 9, 14. Hydrochloric acid; 10, 23. Centrifuge; 11. Waste solvents; 12. polyvinyl alcohol; 13. Acetalization reactor; 15. n-Butyraldehyde;16. Filter; 17 PVB solution; 18. Precipitation kettle; 19. Alkali;20, 22. Water; 21. Washing kettle; 24. Waste water; 25. Dryer; 26. Vibrating screen; 27. PVB.

Figure 6.2: Process flow diagram of one-step dissolution method.

heterogeneous system, which can also be divided into two processes: dissolution and precipitation.

6.4.2.2.1 Two-step dissolution method

PVA is suspended in methanol, and then hydrochloric acid and *n*-butyraldehyde are added to the suspension. The acetalization reaction is carried out at a temperature of 60 °C. As the reaction proceeds, the PVA is gradually dissolved to form a homogeneous solution, then water is added to precipitate the PVB, and the finished product of the PVB can be obtained after neutralization, filtration, water washing, and drying.

6.4.2.2.2 Two-step precipitation method

In aqueous solution, PVA reacts with *n*-butyraldehyde to form PVB. The resultant PVB precipitates, and the reaction mixture changes from homogeneous phase to heterogeneous phase. The process has been used by DuPont for PVB production. Figure 6.3 represents the process flow diagram of two-step precipitation method.

PVA is dissolved in water to form a solution of a concentration between 8% and 10%. Hydrochloric acid and *n*-butyraldehyde are added to the solution with the ratio of PVA: HCl: *n*-butyraldehyde = 100:8:57. The acetalization reaction is carried out at a temperature of 70 °C. As the reaction proceeds, PVB precipitates. After the reaction is finished, the reaction mixture is neutralized and filtrated. The resultant filter cake is washed and dried to obtain a PVB finished product.

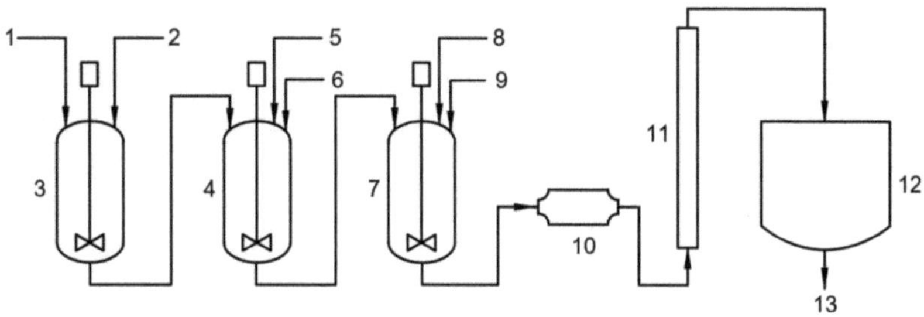

1. Pure water; 2. Polyvinyl alcohol; 3. Dissolution kettle; 4. Condensation kettle; 5. n-Butyraldehyde; 6. Hydrochloric acid; 7. Neutralization kettle; 8. Alkali; 9. Pure water; 10. Filter; 11. Dryer; 12. Finished tank; 13. PVB.

Figure 6.3: Process flow diagram of two-step precipitation method.

In the dissolution method, the reaction system is a homogeneous phase. The reaction is carried out smoothly to reach a high degree of acetalization. The butyral group content in the PVB is in the range from 65% to 80%. Furthermore, the prepared PVB have uniform particle sizes. Other advantages of the dissolution method include little formation of intermolecular acetals, no crosslinking and uniform distribution of acetal groups. However, the separation and purification of the PVB from the solvent are difficult, and the resin is easy coloring and has poor thermal stability because of entrainment of the solvent in PVB particles. In addition, complex solvent recovery and expensive equipment investment are required.

The precipitation method has the advantage that water is used as a solvent and the purity of the prepared PVB resin is high. In addition, organic solvent and corresponding solvent recovery unit are not required, leading to a low production cost. However, in the precipitation method, the acetalization is not uniform, and the intermolecular crosslinking reaction is easy to occur, which affects the solubility of the prepared PVB.

At the same degree of acetalization, the properties of PVB are closely related to the preparation method. A series of PVB resins with different properties can be produced by using different PVAs with different degrees of polymerization and alcoholysis. The density of the resin, the particle structure, the absorption percentage of the solvent and the plasticizer, and even the transparency, can be controlled by adjusting the process conditions.

6.4.3 Uses of PVB resin

The molecular structure of PVB determines that it has high tensile strength and impact resistance as well as good transparency and elasticity. These properties are unmatched by other polymers. PVB is mainly used in safety glass interlayer materials. It is also widely used in special coatings, adhesives, ceramic transfer paper, aluminum foil paper and others [14, 15].

6.4.3.1 Used as a safety glass interlayer

PVB resin has excellent optical clarity and weather resistance, and can maintain no deformation in a wide range of temperatures. It also has rigidity combined with flexibility and excellent impact resistance. Moreover, it has excellent adhesion efficiency with various glass surfaces. Therefore, the high molecular weight PVB resin is added with 30–40% plasticizer and extruded into a film, which can be used as an interlayer material for safety glass. This is the main purpose of PVB. At present, more than 80% of PVB resin in the world is used as interlayer material for safety glass.

Safety glass is a composite glass made of PVB film and two or more layers of glass at high temperature and pressure. When impacted by a foreign object, only irregular radial cracks are formed at the point of impact, and the glass fragments are still firmly adhered to the interlayer without scattering, falling off, causing injury and not affecting the line of sight for a certain period of time. In this case, it can continue to be used with good safety performance. After adding different additives, the light intensity, energy and sound transmission can also be controlled. Therefore, it is widely used in the windows for automobile, aircraft and buildings, special protective glass, navigation, and other aspects.

6.4.3.2 Application on ceramic transfer paper

PVB ceramic decal paper is made by pressing a layer of PVC film on wood paper, coating with PVB film and printing patterns for ceramic printing. PVB decal is easy to use, involves low cost and has a high color burn rate. Since 1985, the demand for PVB in the ceramic decal industry has been increasing, with an annual growth rate of 20%. At present, the amount of PVB consumed by ceramic decals has accounted for 22% of the total output of PVB in China. It has been formed a formal standardized and large-scale production in the ceramic industry of China, but this method is no longer used in other countries.

6.4.3.3 Application in the coating industry

PVB is used in coatings for preparation of primers, baking paints, varnishes, etc. In China, the main products include butyral phosphating primers and modified furan paints. In most cases, PVB is mixed with other resins, and the hydroxyl group in the PVB molecule provides an active group that chemically binds to other thermosetting resins.

PVB is widely used in metal coatings. It is mixed with other components such as phenolic, melamine, epoxy and isocyanate to improve the uniformity of the coating film, increase the adhesive strength and elasticity and minimize the coating layer.

PVB is used as wash primer and sealant for wood products, so that wood products are not deformed, and the coating has strong internal adhesion, moisture resistance, good elasticity, strong rigidity, impact resistance and no color fading. PVB is used for textile coating to improve the waterproofness, coloring and adhesion of fabrics.

PVB is also used in heat-sealing paint, can paint, temporary protective coating for packaging and powder coating.

6.4.3.4 Application in the adhesive industry

PVB-containing adhesives are mainly high-performance structural adhesives, thermosetting and melt adhesives. PVB is an excellent melt adhesive matrix. In formulations with other components, PVB provides toughness, elasticity and high adhesion force, particularly suitable for surfaces that are difficult to bond. The mixture of PVB and phenolic resin is used for bonding metal products, instead of the traditional method of connecting metal products with rivets, so that the joints of the products are more stressed, and the weight of the products is also reduced. Therefore, PVB adhesives are widely used in packaging, bookbinding, labels, metal tin foil and plastic film pasting.

6.4.3.5 Application on aluminum foil paper

With the rapid development of the national economy and export trade, the requirements for product packaging are getting higher and higher. In order to reduce costs, vacuum aluminized paper with PVB coating has emerged and developed rapidly. At present, this is the largest use of PVB resin in China, accounting for 44% of the total PVB output.

6.4.3.6 Other applications

PVB resin can also be applied to prepare special inks such as recording inks, letterpress printing and gravure printing inks. It can also be used to make waterproof fibers, making the fabric antipollution without affecting the appearance, feel and color of the fabric, which is suitable for treatment of almost all fabrics. Furthermore, PVB can be used in dental restorations, wound dressing and wound sutures.

References

[1] Diao C. Technical and economic analysis of different process routes of vinyl acetate in China, Technology & Economics in Petrochemicals 2009, 25(6), 10–13.
[2] Hou L. Production and market analysis of vinyl acetate at home and abroad, Acetaldehyde Acetic Acid Chemical Industry 2009, (8), 9–12.

[3] Liu J. Technical progress and market analysis of vinyl acetate production, Chemical Intermediate 2010, (8), 57–64.
[4] Hu C. Analysis and measures of influencing factors of catalyst activity for vinyl acetate synthesis by acetylene method, Modern Chemical Industry 2006, 26(9), 52–54.
[5] Hou C, Feng L, Li Z, et al. Study on the mechanism of the carboxyl and carbonyl groups on the surface of the catalyst support for the synthesis of vinyl acetate by acetylene, Acta Chimica Sinica 2009, 67(13), 1528–1532.
[6] Chen C, Lin X, Chen X, et al. Research progress of catalysts for the synthesis of vinyl acetate by acetylene, Industrial Catalysis 2003, 11(11), 7–12.
[7] Zhou G, Jiang Y, Lu S, et al. The effect of the structure of the activated carbon support with high specific surface area on the activity of the catalyst for the synthesis of vinyl acetate by acetylene, Petrochemical Technology 2004, 33(7), 608–611.
[8] Liu T, Wang B. Preparation and surface properties of polyvinyl acetate emulsion, Surface Technology 2018, 47(5), 182–187.
[9] Guan L. Industrial production of polyvinyl acetate with gum-based candy, The Food Industry 2017, 38(12), 46–48.
[10] Luo X, Zhang Q, Cheng Y. Research on alcoholysis of high molecular weight polyvinyl acetate, Chinese Journal of Colloid & Polymer 2006, 24(2), 22–23.
[11] Chen T. China's polyvinyl alcohol production technology research progress and market analysis, Acetaldehyde Acetic Acid Chemical Industry 2020, 3, 4–9.
[12] Song Q, Ma Y, Zhao R, et al. New technology for producing polyvinyl butyral by suspension method, China Adhesives 2012, 21(10), 18–21.
[13] Ding L, Liu J, Guo X. Production of polyvinyl butyral by homogeneous alcohol method, Journal of Qingdao University of Science and Technology (Natural Science Edition) 2010, 31(3), 290–293.
[14] Li G, Zhang Y. Research and application of polyethanol butyral resin, China Adhesives 2006, 15(6), 27–32.
[15] Yuan H, Ma P, Wang J, et al. Performance and application of polyvinyl butyral resin, Engineering Plastics Application 2004, 32(2), 43–46.

Chapter 7
Other fine chemicals from acetylene

7.1 Vinylacetylene, chloroprene and neoprene

Vinylacetylene, also known as butenyne, monovinylacetylene, 1-butene-3-yne, 1-butyn-3-ene, 3-buten-1-yne, buten-3-yne, ethynylethene, 1-butenyne and vinylethyne, is a derivative of acetylene. CAS Registry Number: 689-97-4. Molecular formula: C_4H_4, molecular weight: 52.0746. The chemical structure is

Chloroprene, also known as 2-chloro-1,3-butadiene, 2-chlorobutadiene, 2-chlorobuta-1,3-diene, β-chloroprene, chlorobutadiene, 2-chloor-1,3-butadieen, 2-chlor-1,3-butadien, β-chlorobutadiene and 2-chloroprene, is a derivative of vinylacetylene. CAS Registry Number: 126-99-8. Molecular formula: C_4H_5Cl, molecular weight: 88.535. The chemical structure is

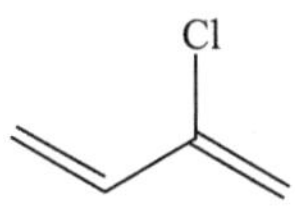

7.1.1 Properties of vinylacetylene, chloroprene and neoprene

Vinylacetylene is a gas with a smell similar to acetylene. The melting point is −118 to −120 °C, the boiling point is 5.5 °C under atmospheric pressure, and the boiling point of vinylacetylene under different pressures is shown in Table 7.1.

Table 7.1: The boiling point of vinylacetylene under different pressures.

Pressure (mmHg)	48.8	107	160	237	320	401	499
Boiling point (°C)	−47	−35	−28	−21	−15	−10	−5
Pressure (mmHg)	621	757	952	1,166	1,380	2,400	4,100
Boiling point (°C)	0	5	12	17.6	22.4	40	60

Chloroprene is a colorless and volatile synthetic liquid that has a pungent ether-like odor. It is soluble in ethanol and diethyl ether, miscible with acetone and benzene, and slightly soluble in water. Other physical properties of chloroprene are given in Table 7.2.

Neoprene, also known as polychloroprene and chloroprene rubber, is an elastomer produced by polymerization of chloroprene. The appearance of neoprene is

https://doi.org/10.1515/9783110714999-007

Table 7.2: Physical properties of chloroprene.

Melting point/°C	−130	Flash point/°C	−20
Boiling point/°C	59.4	Density (20 °C)/g · cm^{-3}	0.958
Vapor pressure/kPa		Refractive index (20 °C)	1.4583
6.4 °C	13.3	Viscosity (20 °C)/mPa · s	0.394
25 °C	28.8	Enthalpy of vaporization (308 K)/kJ · mol^{-1}	30.9
47 °C	66.5	Solubility in water (mg · L^{-1})	875
Henry constant/L · atm · mol^{-1}	56	Octanol–water partition coefficient (log *P*)	2.53

milky white, beige or light brown flakes or lumps. Its density is 1.23–1.25 g cm^{-3}. The solubility parameter δ of neoprene is 9.2–9.41. Neoprene is soluble in toluene, xylene, chloroform, and dichloroethane, and slightly soluble in acetone, methyl ethyl ketone, ethyl acetate, cyclohexane but insoluble in *n*-hexane and solvent gasoline.

Neoprene has good physical, chemical and mechanical properties. In comparison with natural rubber, it is much denser, more resistant to water and hydrocarbon solvents, less permeable to many gases, and was more resistant to degradation by oxygen, ozone, hydrogen chloride, hydrogen fluoride, and other chemicals. It has certain flame retardance. The decomposition temperature is 230–260 °C, and it can be used for a long time at 80–100 °C.

7.1.2 Process for manufacture of vinylacetylene

Acetylene is dimerized in the presence of hydrochloric acid solution of cuprous chloride and ammonium chloride to obtain vinylacetylene.

7.1.2.1 Acetylene dimerization

Acetylene dimerization reaction is usually carried out in a bubbling bed reactor. Acetylene gas is continuously introduced into a liquid solution containing the catalyst, acetylene gas undergoes a dimerization reaction to yield vinylacetylene. The unreacted acetylene gas and the generated vinylacetylene flow out of the reactor together, and are separated in a subsequent separation system to obtain a product containing small amounts of impurities. The unreacted acetylene is returned to the feed system. The reaction equations are as follows:

Dimerization reaction:

$$HC\equiv CH + HC\equiv CH \longrightarrow H_2C=\underset{H}{C}-C\equiv CH$$

Side reaction:

$$HC\equiv CH + H_2C=\underset{H}{C}-C\equiv CH \longrightarrow H_2C=\underset{H}{C}-C\equiv C-\underset{H}{C}=CH_2$$

$$HC\equiv CH+HCl \longrightarrow H_2C=CHCl$$

$$HC\equiv CH+H_2O \longrightarrow CH_3CHO$$

Among them, vinylacetylene is the main product, and there are some by-products including a small amount of divinylacetylene, vinyl chloride, acetaldehyde and a trace of oligomers with high boiling point.

The Nieuwland catalyst has been widely used in the dimerization of acetylene due to its simple preparation and application, mild reaction conditions, good safety and low cost.

7.1.2.2 Acetylene dimerization catalyst

7.1.2.2.1 Nieuwland catalyst

The Nieuwland catalyst is prepared by dissolving the active component cuprous chloride together with a co-solvent in deionized water. It is suggested that the copper chloride complex ions ($CuCl_2^-$, $Cu_2Cl_3^-$, $Cu_3Cl_4^-$, $Cu_4Cl_5^-$) play a catalytic role in the reaction [1–5].

The co-solvent of the catalyst is a chlorine-containing ionic compound. Chlorine ion and cuprous chloride form a complex ion to promote its dissolution. The type and dosage of co-solvent are important factors affecting the catalytic effect. Usually, KCl and NH_4Cl are used as the co-solvent [5, 6]. The free Cl^- in the catalyst solution inhibits the dimerization of acetylene. Therefore, in order to maintain sufficient activity of the Nieuwland catalyst, the amount of the co-solvent should not be too much. The optimal dosage of the co-solvent is just enough to make CuCl complete dissolution in water.

The dimerization of acetylene in the presence of the Nieuwland catalyst requires very strict acidity, and its adjustable range is narrow. The dimerization of acetylene is retarded by hydrogen ion in the catalyst solution, and the higher the concentration of hydrogen ion, the lower the dimerization rate of acetylene. The mass fraction of hydrogen ion in the catalyst system should not exceed 0.1%, otherwise the by-products will increase, and the conversion of acetylene and the selectivity to vinylacetylene will decrease.

The preparation of the Nieuwland catalyst includes a two-step process and a one-step process. In two-step process, copper reacted with chlorine in hydrochloric acid to prepare a wet cuprous chloride with a water content of 23–30%; then, the resultant wet cuprous chloride was mixed with ammonium chloride and other components, and the mixture was heated under stirring to obtain the catalyst solution. The two-step process for preparation of the Nieuwland catalyst was mature, but it took a long time, and the utilization of copper in the preparation of wet cuprous chloride was only 76–81%. The preparation had been improved by one-step process, in which electrolytic copper blocks and hydrochloric acid was added to aqueous ammonium chloride solution, and chlorine was introduced into the solution under stirring. The activity of the catalyst prepared by one-step process was almost the same as that by two-step process, but the preparation time was shortened and the copper was fully utilized in the one-step process [7].

7.1.2.2.2 Nonaqueous phase catalyst for acetylene dimerization

At present, the industrial production of vinylacetylene mainly uses aqueous reaction system, but the conversion of acetylene per pass is only about 10%, and various by-products are also generated. Dimerization of acetylene in nonaqueous phase reaction system was proposed by DuPont Company. In this system, the conversion of acetylene is higher, and no hydration reactions of acetylene or vinylacetylene occur, resulting in less by-products [8].

In the nonaqueous phase catalyst, cuprous chloride is still used as the active component, a primary or secondary aliphatic amine hydrochloride is used as a co-solvent. Cuprous chloride and the co-solvent are dissolved in an organic solvent having excellent solubility for acetylene. Usually, the preferred co-solvents are ethylamine hydrochloride and dimethylamine hydrochloride, while the best solvent is dimethylformamide. For example, in the presence of a catalyst system prepared by dissolving 140 g of cuprous chloride and 70 g of dimethylamine hydrochloride in 150 mL of dimethylformamide, acetylene dimerization reaction was carried out at a temperature of 65 °C and acetylene space velocity of 100 h^{-1}, resulting in acetylene conversion of 20% and the selectivity to vinylacetylene of 92%. In another research, the optimum catalyst solution containing 53 g of CuCl, 25 g of dimethylamine hydrochloride and 54 mL of dimethylformamide was used, and the reaction was performed at a temperature of 67 °C and acetylene space velocity of 320 h^{-1}. The conversion of acetylene and the selectivity to vinylacetylene reached 15.53% and 94.6%, respectively [9].

7.1.2.3 Acetylene dimerization reaction condition

The effects of reaction conditions on the dimerization process of acetylene are mainly reflected in three aspects: reaction temperature, acetylene space velocity and acetylene pressure. In the dimerization of acetylene, the increase of temperature will aggravate the occurrence of side reactions, resulting in the decrease of selectivity of vinylacetylene; However, if the temperature is too low, the reaction rate is very slow, so that although high selectivity can be obtained, the production capacity is too small, which is also not conducive to actual production. Jiao et al. [10] showed that 80 °C was the optimal reaction temperature in aqueous system by orthogonal experiment; In the nonaqueous system with dimethylamine hydrochloride as the promoter, yellow needle-like crystals were formed when the temperature was lower than 65 °C, and insoluble substances were precipitated when the temperature was higher than

70 °C due to the increased volatilization of the solvent, so the reaction temperature was selected as 67 °C [9].

The conversion of acetylene and the yield of vinylacetylene decreased with the increase of acetylene space velocity. In industrial production, two factors, production capacity and conversion per pass of acetylene, must be considered comprehensively.

The effect of acetylene partial pressure on the dimerization rate is related to the composition of the catalyst. The reaction order of acetylene is 1, 1.63 and 2 with the change of acidity and chloride ion concentration in the reaction process [11]. On the premise of ensuring safe production, increasing the partial pressure of acetylene can significantly improve the reaction rate. The acetylene inlet pressure of the reactor in Denka Company Limited (Japan) is higher than that in Chinese manufacturers, which results in a great difference in the production capacity of a single tower (15,000 t a^{-1} in Japan and 7,500 t a^{-1} in China) under the condition that the diameters of reaction towers are similar (2,500 mm in Japan and 2,000–2,200 mm in China).

7.1.2.4 Improvement of acetylene dimerization process and reactor

In general, acetylene dimerization is carried out in a fully mixed bubble column reactor. However, in this type of reactor, the product vinylacetylene is easily retained in the catalyst and further reacts with acetylene to form divinylacetylene and polymers, so the process or reactor must be improved in order to improve the selectivity of vinylacetylene.

Dupont Company suggested that a solvent (ether, higher ketone or higher alcohol) of 1–1.5 times the volume of the Nieuwland catalyst was introduced into the reaction system for exaction of the vinylacetylene generated in the reaction. The reaction system formed two phases including an aqueous phase of the catalyst and an organic phase for dissolution of the vinylacetylene. The acetylene is only dissolved in the aqueous phase of the catalyst, which avoided further contact between the vinylacetylene and the acetylene to generate by-product. A stirred tank reactor was adopted for the reaction. A part of reaction mixture was continuously taken out from the reactor and sent to a separator for separating into two layers. The water layer containing the catalyst was sent back to the reactor, and the organic layer containing the product was subjected to fractional distillation for isolation of vinylacetylene from the solvent. The recovered solvent was recycled to the reactor. The process flow diagram is shown in Figure 7.1.

In the process, the catalyst and the solvent need to be taken out from the reaction area, and excess operation processes of phase separation and fractional distillation were required, increasing the equipment investment and the operation cost. The process could be improved by introduction of an inert organic vapor into the reactor for extraction of the vinylacetylene [12]. Acetone showed the best extraction effect. The reactor using acetone extraction is displayed in Figure 7.2. The method was simple to operate, and that catalyst did not need to be continuously taken out from

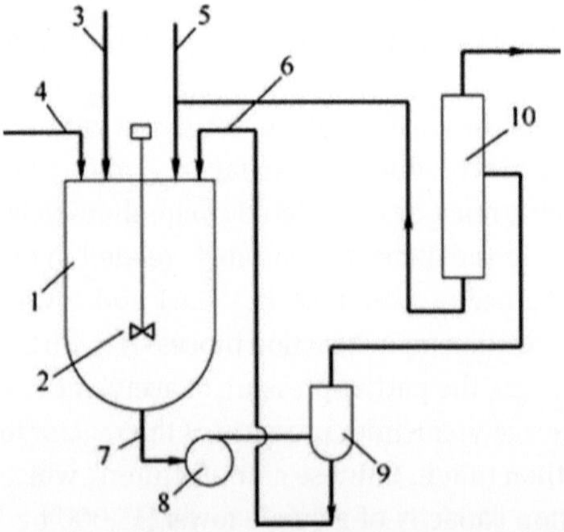

1. Reactor, 2. Stirrer, 3. Fresh catalyst inlet, 4. Acetylene inlet, 7. 5. Solvent, 6. Recycling catalyst inlet, 7. Outlet, 8. Pump, 9. Separator, 10. Fractional distillation tower.

Figure 7.1: Process flow diagram for dimerization of acetylene.

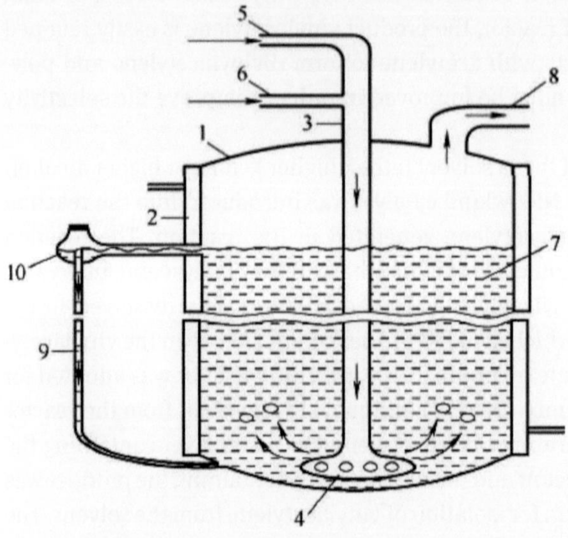

1. Reactor, 2. Heating jacket, 3. Main gas inlet, 4. Gas dispersion head, 5. Acetylene inlet, 6. Acetone inlet, 7. Catalyst solution, 8. Gas outlet, 9. Circulating pipe, 10. Intermediate collector.

Figure 7.2: Acetylene dimerization reactor using acetone stripping.

the reaction zone, and simultaneously the vinylacetylene was extracted in time to avoid further reaction with acetylene.

Zhan et al. [13] investigated the effects of four bubble column reactors on the dimerization of acetylene. In comparison with standard bubble column, using the external circulation reactor and the packed reactor increased the conversions of acetylene by 5.15% and 9.65%, and the selectivity to vinylacetylene by 4.85% and 14.85%, respectively. When the internal loop reactor was used, the conversion of acetylene was almost unchanged, but the selectivity to vinylacetylene decreased by 2.8%. Obviously, from the effect of acetylene dimerization reaction, the packed reactor was the best among the four bubble reactors, and next was the external circulation reactor. However, the packed bubbling bed reactor was easy to be blocked by polymer and the resistance was large, so it was not suitable to use in actual production. Therefore, the external circulation reactor was the most suitable reactor for acetylene dimerization. The different reaction effects obtained by using the external circulation reactor and the internal loop reactor were mainly caused by the different back mixing conditions of the gas phase in the two types of reactors. In the external circulation reactor, as the catalyst of the external circulation flow returned to the bottom of the reactor and forms an ascending liquid flow, the back mixing of a gas phase can be obviously reduced, increasing the average concentration of reactants in the catalyst solution, and thus accelerating the reaction rate of acetylene dimerization. Moreover, the acetylene further reaction with vinylacetylene was decreased. Therefore, both the conversion of acetylene and the selectivity to vinylacetylene were improved. However, in the internal loop reactor, the continuous circulation of the catalyst in the reactor enhances the degree of gas back mixing, thereby increasing the average concentration of the product in the reactor and facilitating the further reaction between the product and acetylene.

7.1.3 Process for manufacture of chloroprene

At present, there are two industrial processes for production of chloroprene in the world: acetylene route and butadiene route. In China, all the chloroprene is produced by acetylene route, yet butadiene route dominates the production of chloroprene in other countries [14].

The production of chloroprene by acetylene route includes two basic processes: synthesis of vinylacetylene from acetylene and synthesis of chloroprene from vinylacetylene. In the presence of catalyst solution composed of cuprous chloride and an alkali metal salt, acetylene undergoes dimerization reaction to produce vinylacetylene and various by-products such as divinylacetylene, acetaldehyde and vinyl chloride. Pure vinylacetylene is obtained by absorption, desorption and fractional distillation.

Vinylacetylene undergoes an addition reaction with hydrogen chloride in the presence of cuprous chloride catalyst in hydrochloric acid solution to yield chloroprene:

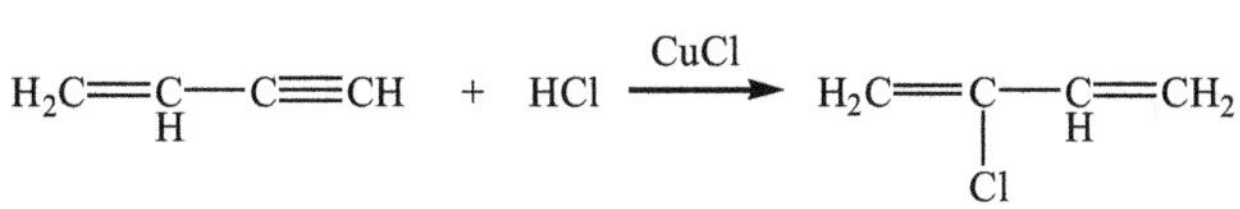

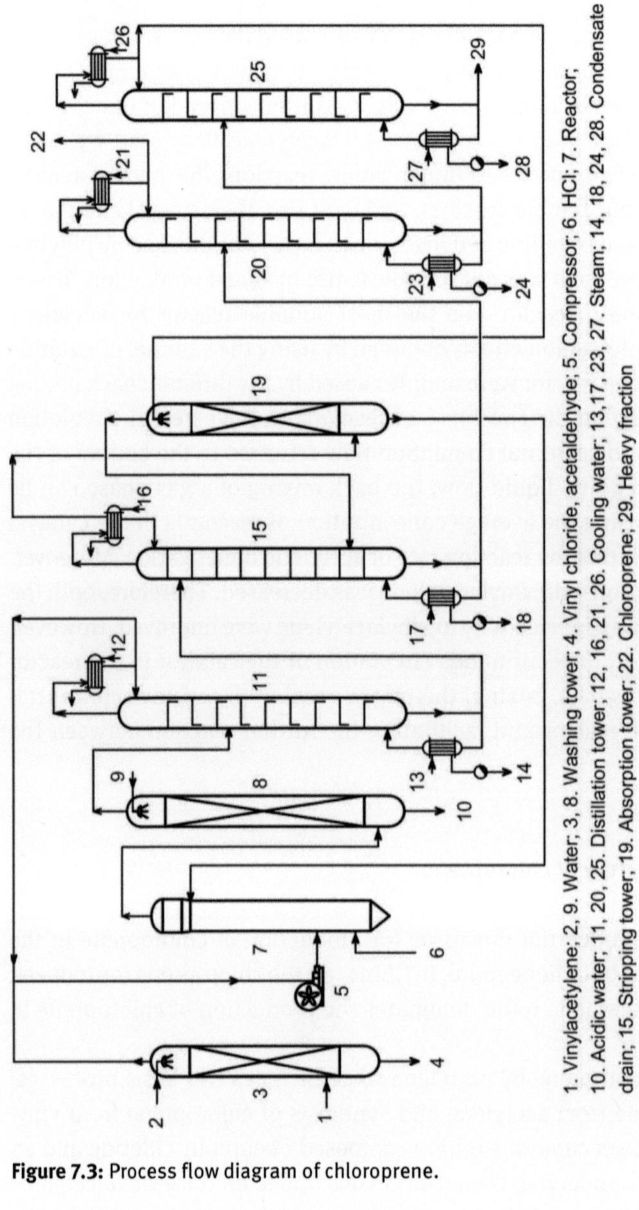

Figure 7.3: Process flow diagram of chloroprene.

Some by-products such as methyl vinyl ketone and 1-chloro-1,3-butadiene are also generated. The unreacted vinylacetylene is recovered by fractional distillation and recycled. The crude product is cooled and refined to obtain the polymerization-grade chloroprene.

The process flow diagram for production of chloroprene is shown in Figure 7.3. Vinylacetylene was washed by water in the washing tower and introduced into the reactor filled with the catalyst solution by a compressor, and hydrogen chloride is simultaneously introduced into the reactor. The reaction was carried out at a temperature of 40 to 45 °C. The reaction mixture was discharged from the top of the reactor and introduced into the fractional distillation tower. The top product of the tower was the unreacted vinylacetylene, which returned to the reactor. The bottom product of the tower was the crude chloroprene, which was subjected to stripping, absorption and fractional distillation to obtain the finished product of chloroprene.

Denka Company Limited (Japan) has advanced technology of acetylene route to chloroprene. The yield of chloroprene is 95% based on vinylacetylene and 82% based on acetylene. The conversion of acetylene dimerization per pass is 20% and the yield of vinylacetylene is 88%. Each ton of chloroprene consumes 2.16 tons of calcium carbide and 6 tons of steam.

7.1.4 Uses of vinylacetylene, chloroprene and neoprene

Vinylacetylene is mainly used for the production of chloroprene, leaf alcohol and methanol adhesives. It is also used for the production of divinyl ether and methyl vinyl (methyl) ketone-based polymer.

Chloroprene is mainly used for production of neoprene by free radical emulsion polymerization. The polymerization of chloroprene is initiated using potassium persulfate. Zinc oxide and thioureas are used for crosslinking of individual polymer strands.

7.1.4.1 Neoprene

Neoprene has good comprehensive performance. It possesses excellent physical and mechanical properties, and the tensile strength is similar to that of natural rubber. Because neoprene molecular chain contains polar chlorine atom groups, on the one hand, the double chain is protected to weaken the activity, and on the other hand, the polymer has great stability to nonpolar substances, so the light resistance, heat resistance, aging resistance, oil resistance and chemical corrosion resistance of neoprene are superior to those of natural rubber, and the ozone aging resistance, weather resistance and flame resistance are more outstanding.

Neoprene is widely used in the manufacture of industrial rubber products such as wire and cable sheaths, oil-resistant hoses and rubber plates, conveyor belts, rubberized fabrics and various sealing ring gaskets. It is also used for anticorrosive linings of chemical equipment and footwear adhesives. In addition, it has been widely used in the field of construction, electronics industry and modification of other polymer materials.

In the United States, the European Union and Japan, neoprene is widely used in the production of automotive parts. For example, the consumption of neoprene in automotive parts accounts for 50% of the total consumption in Japan.

In China, adhesives used in shoemaking and construction are the largest application fields of neoprene, accounting for 60% of the total consumption, followed by industrial rubber products, accounting for 30% of the total consumption, and another 10% are used in wire and cable and other fields. With the rapid development of the automobile industry, the demand for neoprene is increasing.

7.1.4.2 Methanol adhesive

Methanol adhesive, also known as carbinol adhesive, is a gelatinous resin obtained by polymerization of dimethyl vinylethynyl methanol (DMVEM) monomer under the initiation of benzoyl peroxide. DMVEM is a colorless liquid with a boiling point of 50–51 °C under a pressure of 9–10 mmHg. The refractive index is in the range from 1.4750 to 1.4765 at 20 °C. Usually, a small amount of polymerization inhibitor is added to the monomer to prevent it from self-polymerization during storage.

Methanol adhesive is characterized by high tensile strength (about 200 kg cm^{-2}), excellent oil resistance or cold resistance; especially it can be bonded at room temperature, so it has been widely used in aviation industry, optical industry and machinery manufacturing industry. However, it also has some shortcomings, mainly that it is poor in resistance to water and organic solvent, and is not resistant to high temperatures. It is not ideal under the conditions of bearing shear stress and bending stress.

DMVEM is produced by a condensation reaction of vinylacetylene and acetone in the presence of potassium hydroxide powder catalyst:

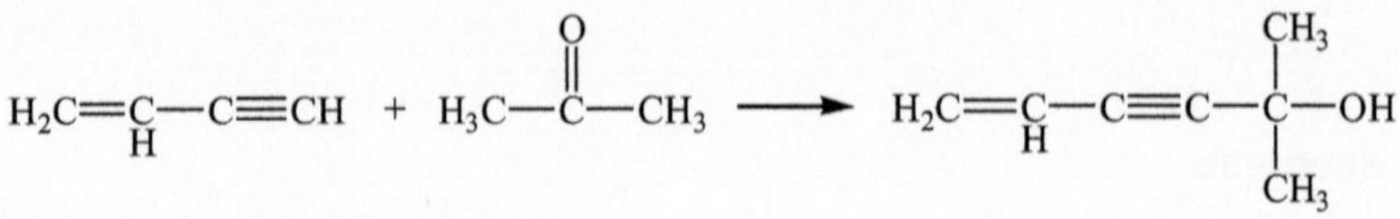

The reaction mechanism is as follows:

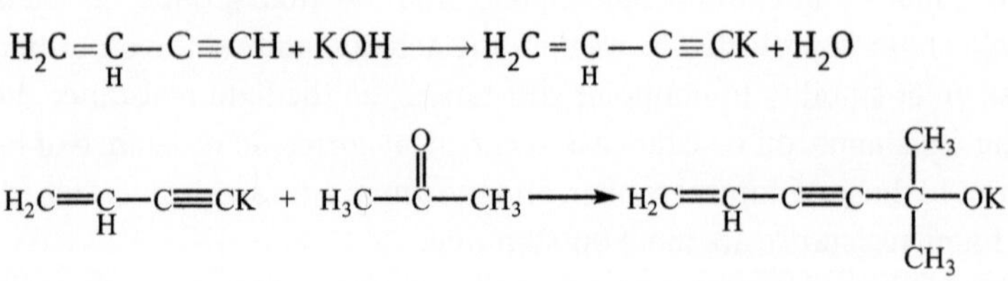

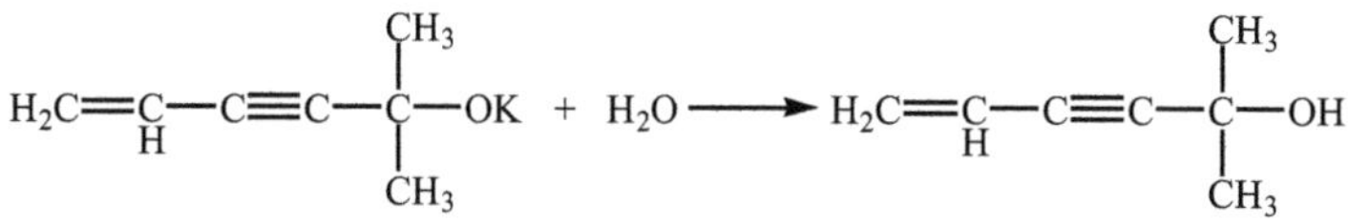

The preparation of the monomer was carried out under strictly anhydrous condition. 60 g of dried potassium hydroxide powder was dispersed in 100 mL of anhydrous diethyl ether. The suspension was cooled to 0 °C, and 60 g of vinylacetylene was added to the suspension. Subsequently, 60 g of acetone was added in a dropwise manner, and the dropping speed was controlled so that the reaction temperature did not exceed 5 °C. After the addition of acetone was finished, the reaction was continued at room temperature for 4 h under vigorous agitation. The resultant potassium dimethyl vinylethynyl methoxide was hydrolyzed at room temperature for 30 min under stirring by adding 100 mL of water to the reaction mixture. After hydrolysis, the reaction mixture was transferred to a separating funnel for two layers. The upper layer of ether solution was collected, washed with 5% sulfuric acid solution and then separated. A small amount of potassium carbonate was added to the ether solution to neutralize the residual sulfuric acid to neutrality. Hydroquinone dosage of 0.1% based on the mass of DMVEM was added as polymerization inhibitor. The solution was distilled under atmospheric pressure for removal of the unreacted reactants and ether, and subjected to distillation under reduced pressure. DMVEM was obtained by collecting the fraction of a boiling point between 50 and 51 °C under 9–10 mmHg. The yield of DMVEM is 88.1% [15].

7.2 Leaf alcohol

Leaf alcohol, also known as (*Z*)-hex-3-en-1-ol, *cis*-hex-3-en-1-ol, *cis*-3-hexen-1-ol, *cis*-3-hexene-1-ol, *cis*-3-hexenol, 3-(*Z*)-hexenol; 3-(*Z*)-hexen-1-ol, (*Z*)-3-hexen-1-ol, is a derivative of acetylene and ethylene oxide. CAS Registry Number: 928-96-1. Molecular formula: $C_6H_{12}O$, molecular weight: 100.1589. The chemical structure is

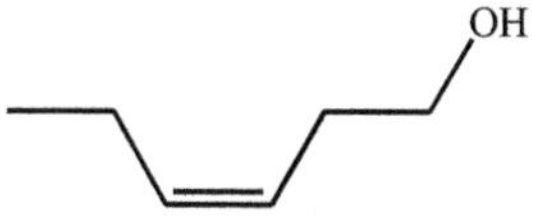

7.2.1 Properties of leaf alcohol

Leaf alcohol is colorless oily liquid with boiling point of 156–157 °C, density of 0.8460 g cm^{-3} and refractive index of 1.4395 at 20 °C. It is miscible with most oils and soluble in alcohol and most organic solvents, but only slightly soluble in water. Leaf alcohol has a pleasant and aroma of fresh green grass. It is a representative fragrance with green aroma. Natural leaf alcohol exists in almost all green plants, mainly in tea, mint, jasmine, *Osmanthus*, *Acacia*, mulberry, grapefruit, and so on. The isomer of leaf alcohol is *trans*-hex-2-en-1-ol. The specific fragrant properties of the two isomers are caused by the presence of a double bond and geometric structures on their molecules wherein *cis*-hex-3-en-1-ol possesses a very pleasant fragrance while *trans*-hex-2-en-1-ol lacks typical green odor but has fruity odor [16].

7.2.2 Process for manufacture of leaf alcohol

Leaf alcohol can be isolated from the essential oils extracted from various plants, but only in limited quantities. Commercial leaf alcohol is mainly obtained by chemical synthesis. According to the starting material, the main methods for synthesis of leaf alcohol include acetylene route, vinylacetylene route, butyne route and propargyl alcohol route. The butyne route follows exactly the same procedure as the acetylene route after 1-butyne has been prepared.

7.2.2.1 Acetylene route

In liquid ammonia, acetylene reacts with metal sodium to form sodium acetylide, and then sodium acetylide reacts with bromoethane to form 1-butyne. Further reaction of 1-butyne with sodium in liquid ammonia gives butynyl sodium, which is then reacted with ethylene oxide and hydrolyzed to yield 3-hexyne-1-ol. Partial hydrogenation of 3-hexyne-1-ol produces leaf alcohol.

$$HC{\equiv}CH + Na \xrightarrow{NH_3} HC{\equiv}CNa$$

$$HC{\equiv}CNa + C_2H_5Br \longrightarrow CH_3CH_2C{\equiv}CH + NaBr$$

$$CH_3CH_2C{\equiv}CH + Na \xrightarrow{NH_3} CH_3CH_2C{\equiv}CNa$$

$$CH_3CH_2C{\equiv}CCH_2CH_2 + H_2 \xrightarrow{Pd/CaCO_3} CH_3CH_2CH{=}CHCH_2CH_2OH$$

Acetylene was introduced into the liquid ammonia for the formation of a saturated solution at a temperature of –34 °C. About 5.8 g of sodium was added to the solution, resulting in the generation of sodium acetylide. About 17 mL of bromoethane and

30 mL of anhydrous ether were added, and reaction was carried out for 4 h under agitation. The resultant butyne in the reaction solution was converted into sodium butylide by addition of sodium (5.8 g) and anhydrous ferric nitrate (0.1 g). The reaction solution was cooled to −40 °C, and stirred for 24 h after 23 mL of ethylene oxide was added. Water and 17.3 g of NH_4Cl was added to the reaction solution for hydrolysis. After phase separation, the organic phase was distilled to obtain 3-hexyne-1-ol. Partial hydrogenation of 3-hexyn-1-ol in the presence of Lindlar catalyst ($Pd/CaCO_3$) gave leaf alcohol. The yield is 80%. This is a mature synthetic route, which has the advantages of easily available raw materials, high yield, less by-products and so on, but the disadvantages are complex process and requirement of cryogenics refrigeration.

The reaction of triethylaluminum with acetylene can also give the cis-configuration of the double bond, which can also be used for the synthesis of leaf alcohol [17].

7.2.2.2 Vinylacetylene route

In liquid ammonia, vinylacetylene reacts with metal sodium to form sodium vinylacetylene, and then sodium vinylacetylene reacts with ethylene oxide to give 3-hexyne-5-en-1-ol, and partial hydrogenation of 3-hexyne-5-en-1-ol produces leaf alcohol.

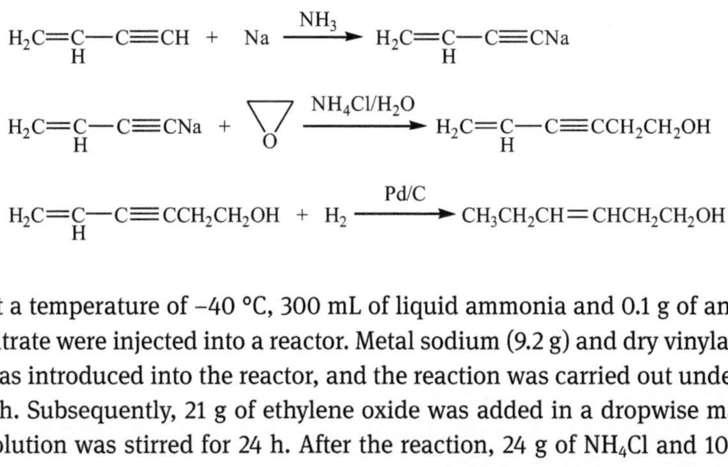

At a temperature of −40 °C, 300 mL of liquid ammonia and 0.1 g of anhydrous ferric nitrate were injected into a reactor. Metal sodium (9.2 g) and dry vinylacetylene (21 g) was introduced into the reactor, and the reaction was carried out under agitation for 2 h. Subsequently, 21 g of ethylene oxide was added in a dropwise manner and the solution was stirred for 24 h. After the reaction, 24 g of NH_4Cl and 100 mL of water were added for hydrolysis. After phase separation, the organic phase was distilled to obtain 21 g of 3-hexyne-5-en-1-ol, which was dissolved in 50 mL of methanol and partial hydrogenated in the presence of 5% Pd/C catalyst at a temperature of 30 °C under a pressure of 1.27 MPa. Distillation of the reaction solution gave 19 g of leaf alcohol with a purity of 91%.

This process had some disadvantages including long reaction time and requirement of cryogenics refrigeration. It was improved by using Grignard reagent. Vinylacetylene was added to Grignard reagent at a temperature of 0 to 10 °C. Subsequently, ethylene oxide was added, and the reaction solution was under agitation for a certain time. Then NH_4Cl and water were added for hydrolysis. 3-Hexyn-5-en-1-ol was obtained with a yield more than 40%. Hydrogenation of 3-hexyn-5-en-1-ol was carried out in the presence of nickel catalyst at room temperature under a pressure of 4 MPa to obtain leaf alcohol with the purity of more than 60%. The improved synthetic route did not use liquid ammonia as solvent, so that the reaction became mild, the operation time was greatly shortened and the yield of leaf alcohol was increased. The improved process was more suitable for industrial production of leaf alcohol.

7.2.2.3 Propargyl alcohol route

Propargyl alcohol was converted to propyne chloride. The latter reacted with methyl Grignard reagent to obtain butynyl magnesium chloride, which was then reacted with ethylene oxide and hydrolyzed to give 3-hexyne-1-ol. Partial hydrogenation of 3-hexyne-1-ol in the presence of palladium catalyst yields leaf alcohol. The purity of leaf alcohol was up to 99% and the overall yield was 37.8%.

The propyne chloride may be replaced with *p*-toluenesulfonyl propynyl ester under substantially the same reaction conditions. The yield of leaf alcohol is slightly improved and the purity is almost the same.

7.2.3 Uses of leaf alcohol

At present, hundreds of tons of leaf alcohol are consumed in the world, of which 35% is directly used for preparation of essence and 65% is used for production of esters, mainly formate, acetate, benzoate and salicylate of leaf alcohol.

7.2.3.1 Application of leaf alcohol in food

In Japan, leaf alcohol is widely used in the preparation of natural fresh flavor essences such as banana, strawberry, citrus, rose and apple. It is also used together with ester flavors such as acetate, valerate and lactate to improve the taste of food, mainly for inhibiting the sweet aftertaste of cool drinks and fruit juices. In the national standard of food additive of China (GB2760-1996), leaf alcohol is specified to be used in food flavor according to the production needs.

7.2.3.2 Application of leaf alcohol in cosmetics industry

Leaf alcohol possess an intensely green grassy odor, which is a popular type of delicate fragrance. Essences prepared with leaf alcohol are also used in cosmetics in large quantities. More than 40 famous essence formulations in the world contain leaf alcohol. Usually, addition of only 0.5% or less of leaf alcohol can obtain obvious aroma of fresh green leaves. It is used as component of various artificial essential oils similar to natural fragrances such as *Convallaria*, clove, mint and lavender essential oils. It is also used for preparation of various flower essences, giving the artificial essential oils and essences pleasant odors of green leaves.

7.2.3.3 Application of leaf alcohol in biological control

Leaf alcohol is also an indispensable physiological active substance in both plants and insects. Insects use leaf alcohols as pheromones or sex hormones for alarm and assembly. Therefore, leaf alcohol can be used in biological control. For example, the mixture of leaf alcohol and benzoquinone in a certain proportion can induce the aggregation of male scarab and beetle, so as to kill such forest pests in a large area.

7.3 2-Methylpyridine and 2-vinylpyridine

2-Methylpyridine, also known as 2-picoline, α-methylpyridine and α-picoline, is a derivative of acetylene and ammonia. CAS Registry Number: 109-06-8. Molecular formula: C_6H_7N, molecular weight: 93.1265. The chemical structure is

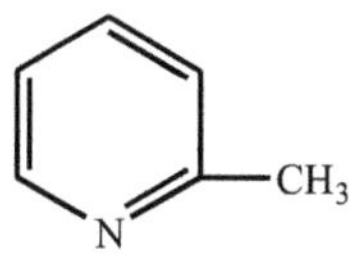

2-Vinylpyridine, also known as 2-ethenylpyridine, α-vinylpyridine, is a derivative of acetylene and acrylonitrile. CAS Registry Number: 100-69-6. Molecular formula: C_7H_7N, molecular weight: 105.1372. The chemical structure is

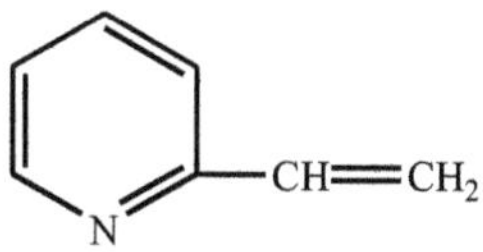

7.3.1 Properties of 2-methylpyridine

2-Methylpyridine is a colorless and oily liquid with an odor of pyridine. Its melting point is −67 °C and boiling point is 129 °C. It is soluble in water and miscible with alcohol and ether. 2-Methylpyridine forms an azeotrope with water at 92.8 °C and the mass fraction of 52%. The physical properties of 2-methylpyridine are listed in Table 7.3.

Similar to pyridine, 2-methylpyridine can react with inorganic or organic acids to form salts. The methyl group in 2-methylpyridine is reactive and forms pyridine-2-carboxylic acid upon oxidation. Hydrogenation of 2-methylpyridine leads to 2-methylpiperidine.

Table 7.3: Physical properties of 2-methylpyridine.

Melting point/°C	−67	Flash point (open cup)/°C	29
Boiling point (101.3 kPa)/°C	129	Enthalpy of combustion of liquid/ kJ · mol^{-1}	3,420
Density (20 °C)/g · cm^{-3}	0.9455	Enthalpy of vaporization/kJ · mol^{-1}	42.94
Refractive index (20 °C)	1.5020	Enthalpy of fusion/kJ · mol^{-1}	9.72
Dielectric constant (22 °C)	9.46	Heat capacity (25 °C) /J · mol^{-1} · K^{-1}	158.41
Dipole moment/ × 10^{-30}C · m	6.54	Enthalpy of formation of liquid/ kJ · mol^{-1}	58.99
Viscosity (20 °C)/mPa · s	0.8102	Critical temperature/°C	348
Surface tension (20 °C)/N · m^{-1}	33.18 × 10^{-3}	Critical pressure/MPa	4.6

2-Methylpyridine reacts with paraformaldehyde at 200 °C resulting in the formation of 2-(β-hydroxyethyl) pyridine.

7.3.2 Properties of 2-vinylpyridine

2-Vinylpyridine is a colorless liquid with a boiling point of 159–160 °C and a flash point of 46 °C. The vapor pressures are 0.5 kPa at 52.5 °C and 3.9 kPa at 80.5 °C. The density is 0.9885 g cm^{-3} at 20 °C. The refractive index is 1.5495 at 20 °C. It is slightly soluble in water, but easily soluble in ethanol, ether, chloroform, benzene, and acetone. It is lachrymatory and toxic. Polymerization of 2-vinylpyridine is liable to occur under light or heat, therefore, 4-*tert*-butyl catechol of 0.1% is usually added to 2-vinylpyridine during storage to prevent it from polymerization.

7.3.3 Process for manufacture of 2-methylpyridine

At present, there are mainly four processes for production of 2-methylpyridine: acetylene–ammonia route, acetylene–acetonitrile route, acetaldehyde–ammonia route, and separation from coal tar.

7.3.3.1 Manufacture of 2-methylpyridine from acetylene and ammonia (acetylene–ammonia route)

In the presence of a catalyst, acetylene reacts with ammonia at a temperature of 400–500 °C to produce 2-methylpyridine and 4-methylpyridine:

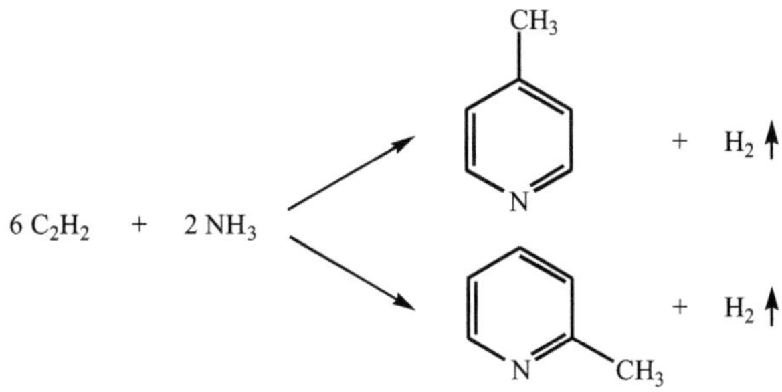

A side reaction occurs at the same time and leads to a by-product acetonitrile:

$$C_2H_2 + NH_3 \longrightarrow CH_3CN + H_2$$

For the reaction of acetylene and ammonia, the change of the catalyst has a great influence on the selectivity of the reaction. Using $CdSO_4$–$A1_2O_3$ as the catalyst, the solution containing 24.2% of 2-methylpyridine, 12% of 4-methylpyridine and a small amount of by-product acetonitrile was obtained at a reaction temperature of 350 °C and C_2H_2/NH_3 molar ratio of 1.5:1. In the presence of $Cd_3(PO_4)_2$-Al_2O_3 catalyst, the conversion of acetylene was 65–80% and the selectivities to 2-methylpyridine and 4-methylpyridine were 59.6% and 32.4% at a reaction temperature of 420–440 °C and the C_2H_2/NH_3 molar ratio of 1:2. When $Cd_3(PO_4)_2$ supported on bleaching soil was used as a catalyst and acetylene reacted with ammonia at a temperature of 300 °C, the condensed product contained 50% of 2-methylpyridine, while the acetonitrile by-product was few. When an oxide such as CdO, ZnO, $A1_2O_3$ or Cr_2O_3 is added to the catalyst, and the temperature is raised to 300–400 °C, the amount of by-produced acetonitrile is increased. With the CdO-bleaching clay catalyst, the reaction product is mainly 2-methylpyridine, but when a small amount of Cr_2O_3, MoO_3 or WO_3 cocatalyst is simultaneously added with a V_2O_5 pumice catalyst, a good pyridine yield can be obtained at a reaction temperature of 350 °C. Using ZnO 50%, kaolin 25% and activated clay 25% as catalysts and the reaction was carried out at 400 °C, the main product was acetonitrile with a yield of 75.1%. Using ZnO–kaolin (1:4) as a catalyst and the reaction was carried out with C_2H_2/NH_3 molar ratio of 1:1 at a temperature of 400 °C, the resultant product containing 91% acetonitrile, and the pyridine base content was less than 0.8%.

7.3.3.2 Manufacture of 2-methylpyridine from acetylene and acetonitrile (acetylene–acetonitrile route)

Using bis(cyclopentadienyl)cobalt as a catalyst, acetylene and acetonitrile are cyclized in toluene at 180 °C and 1.1–1.2 MPa to form 2-methylpyridine. This is currently the most advanced process for production of 2-methylpyridine, with high yield, less by-products and no wastewater. The catalyst bis(cyclopentadienyl)cobalt is obtained by reaction of cyclopentadiene, sodium amide and cobalt chloride. Since water can deactivate the catalyst, the introduction of water must be strictly limited during the preparation of the catalyst, while the moisture contents in the raw materials (acetylene and acetonitrile) should be strictly controlled before the reaction [18].

7.3.3.3 Manufacture of 2-methylpyridine from acetaldehyde and ammonia (acetaldehyde–ammonia route)

Acetaldehyde and ammonia are preheated and then introduced into the catalytic reactor. The reaction is carried out at a temperature of 350–550 °C under atmospheric pressure. Al_2O_3 is used as a catalyst, and some other metal oxides are used as cocatalysts. The outlet gas mixture is condensed, dehydrated and subjected to fractional distillation, leading to 2-methylpyridine and 4-methylpyridine with a total content between 99.2% and 99.5%. The yield is in the range from 40% to 60%, in which each of the two isomers accounts for about one half.

7.3.3.4 Separation from coal tar

The crude 2-methylpyridine obtained by separation from coal tar is subjected to deslagging to obtain a mixture 2-methylpyridine and water, and then the mixture is subjected to distillation in a packed column under atmospheric pressure. Benzene is added for azeotropic distillation to remove the water in it, followed by fractional distillation to obtain 2-methylpyridine.

7.3.4 Processes for manufacture of 2-vinylpyridine

There are two main production routes for 2-vinylpyridine, one is based on 2-methylpyridine and formaldehyde, and the other is based on acetylene and acrylonitrile.

7.3.4.1 Production of 2-vinylpyridine from 2-methylpyridine and formaldehyde

2-Methylpyridine and formaldehyde undergo an addition reaction to form 2-hydroxyethylpyridine, which is then dehydrated to form 2-vinylpyridine. The reaction equations are as follows:

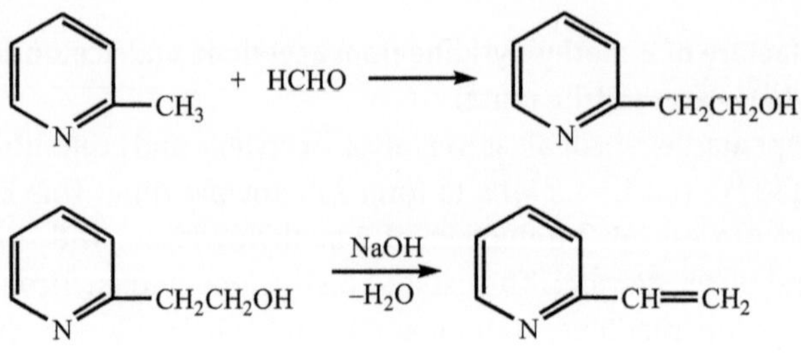

2-Methylpyridine and formaldehyde were mixed in the molar ratio of 1:0.4 and introduced into a tubular reactor by a metering pump. The flow rate was controlled so that the residence time of the reaction solution was 8 min. The reaction was carried out at a temperature of 250 °C under a pressure of 9.08 MPa. 2-Hydroxyethylpyridine was obtained in a single pass yield of 30%. Unreacted 2-methylpyridine was recovered from the reaction solution by washing with 50% sodium hydroxide solution and subsequently undergoing fractional distillation. The recovered 2-methylpyridine was recycled. A sodium hydroxide solution with mass concentration of 50% was added to the remaining liquid after the unreacted 2-methylpyridine had been separated off, and the resultant solution was being heated and refluxed for dehydration. The dehydrated reaction solution was distilled under reduced pressure. Fraction of above 70 °C (8.0 kPa) was collected to obtain crude 2-vinyl pyridine, and then it was purified by fractional distillation to give a finished product 2-vinylpyridine with the content of more than 98%.

7.3.4.2 Production of 2-vinylpyridine from acetylene and acrylonitrile

Acetylene reacts with acrylonitrile in the presence a soluble organic cobalt salt catalyst to produce 2-vinylpyridine:

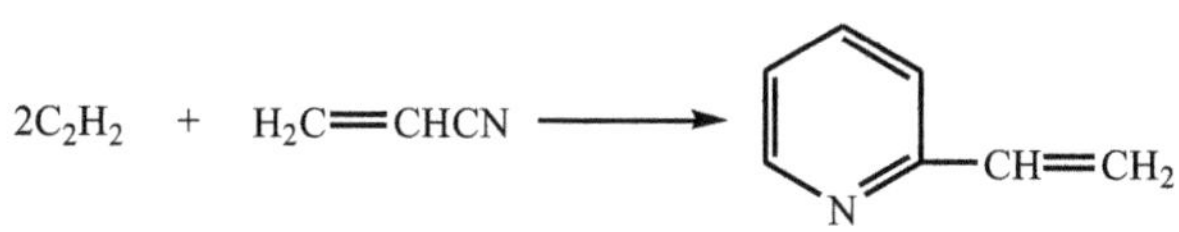

The reaction was carried out at a temperature of 150–160 °C under a pressure of 0.7 to 0.8 MPa for 0.5 h. The yield of 2-vinylpyridine was 93% based on acrylonitrile [19].

7.3.5 Uses of 2-methylpyridine and 2-vinylpyridine

2-Methylpyridine can be used to synthesize an important polymer monomer, 2-vinylpyridine, which is widely used in the rubber industry by homopolymerization or copolymerization with other materials to form a latex.

2-Methylpyridine is also a raw material for synthesis of a novel herbicide 4-amino-3,5,6-trichloropyridine, and it is also a raw material for manufacture of a nitrogen fertilizer synergist 2-chloro-6-(trichloromethyl)pyridine. The nitrogen fertilizer synergist can inhibit nitrifying bacteria in soil, reduce the transformation and loss of ammonium ions, thereby improving the utilization of fertilizers and increasing the yield of crops.

In addition, 2-methylpyridine is used in the pharmaceutical industry as an intermediate for the manufacture of long-acting sulfonamides and the organophosphorus antidote pralidoxime chloride.

2-Vinylpyridine is mainly used to form a terpolymer butadiene–styrene–vinyl pyridine latex with butadiene and styrene. Due to the introduction of a highly polar pyridine group in the molecule, the latex has a strong adhesive ability. It is mainly used for impregnating rayon, nylon, polyester and other cords or cord fabrics. It is widely used in high-strength tires such as airplanes and automobiles, pressure-resistant hoses, conveyor belts, hiking boots and other rubber products. Compared with natural latex or styrene–butadiene latex, butadiene–styrene–vinylpyridine latex can increase the adhesion strength of rayon cord to rubber by 17–67%, and improve the adhesion strength of nylon cord to rubber by 43–122%. In recent years, with the rapid development of the national economy, the output of tire-based rubber products has increased substantially, and the annual growth rate has exceeded 10% in China. The rapid growth in the production of tires and cord fabrics has led to a strong increase in the demand for butadiene–styrene–vinylpyridine latex, leading to a rapid increase in the consumption of 2-vinylpyridine.

In the pharmaceutical industry, the use of a copolymer of 2-vinylpyridine and styrene as a film-coating material for tablets has many advantages and is gradually replacing the conventional sucrose coating material. The coating material can overcome the shortcomings of the sugar coating, such as easy moisture deterioration and cracking, thereby ensuring the stability of the tablet for a long time. 2-Vinylpyridine can also be used in the production of drugs such as polyvinylpyridine oxidum and betahistine hydrochloride.

In the pesticide industry, 2-vinylpyridine is used for synthesis of the herbicide paraquat, and it can also be applied for synthesis of the insecticide imidacloprid with methanol. These pesticide products are the leading varieties of pesticides on the market at present.

7.4 Acrylic acid

Acrylic acid, also known as acroleic acid, ethylenecarboxylic acid, propenoic acid, vinylformic acid, propene acid, and glacial acrylic acid, is a derivative of acetylene and carbon monoxide. CAS Registry Number: 79-10-7. Molecular formula: $C_3H_4O_2$, molecular weight: 72.0627. It is the simplest unsaturated carboxylic acid, consisting of a vinyl group connected directly to a carboxylic acid terminus. The chemical structure is

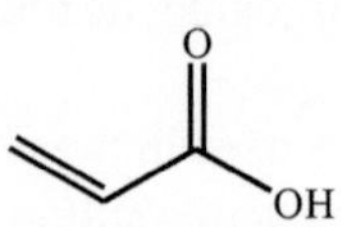

7.4.1 Properties of acrylic acid

Acrylic acid is a colorless liquid with a characteristic acrid or tart smell. It is miscible with water, alcohols, ethers and chloroform. The physical properties of acrylic acid are listed in Table 7.4.

Table 7.4: Physical properties of acrylic acid.

Melting point/°C	13.0	Flash point/°C	54
Boiling point/°C	141.6	Enthalpy of fusion (285.7 K)/kJ · mol^{-1}	9.51
Vapor pressure (20 °C)/mmHg	4	Heat capacity (20 °C)/J · mol^{-1} · K^{-1}	144.2
Density (20 °C)/g · cm^{-3}	1.0511	Enthalpy of vaporization (356 K)/ kJ · mol^{-1}	45.3
Refractive index (20 °C)	1.4210	pK_a (25 °C)	4.25

Acrylic acid is a strong organic acid and thus corrosive. It is chemically active and easily polymerized into a transparent white powder. When it is reduced, propionic acid is formed. When it is added with hydrochloric acid, 2-chloropropionic acid is formed. Acrylic acid can undergo a characteristic reaction of a carboxylic acid, and a corresponding ester can be obtained by reacting with an alcohol. After the acrylic acid and its esters are themselves or mixed with other monomers, polymerization occurs to form a homopolymer or a copolymer. The monomers are usually copolymerized with acrylic acid including amides, acrylonitrile, vinyl-containing styrene and butadiene.

Usually, methoxyhydroquinone or hydroquinone is added to acrylic acid as a polymerization inhibitor.

7.4.2 Process for manufacture of acrylic acid

In the 1960s, the industrial production of acrylic acid was mainly by carbonylation of acetylene (Reppe process). Afterward, with the rapid development of petrochemical industry, the two-step process of direct oxidation of propylene to acrylic acid gradually replaced the Reppe process and became the mainstream process for production of acrylic acid. In 1995, BASF closed down the last unit for production of acrylic acid by Reppe process in Ludwigshafen, Germany, and in 1999, Ciba Specialty Products Company closed down the only unit for production of acrylic acid by acrylonitrile process in Bradford, United Kingdom. Therefore, propylene oxidation process is currently the only way to produce acrylic acid on a large scale. However, Reppe process has attracted renewed interest in recent years as an alternative to the petroleum route. China is poor in oil and rich in coal, and the price of coal is relatively cheap. Acetylene is mainly from calcium carbide, which is an important coal chemical product. A large amount of carbon monoxide is by-produced in the production of calcium carbide. The synthesis of acrylic acid using acetylene and by-product carbon monoxide as raw materials is not only an atomic economic reaction, which is of great significance in the full use of resources, but also an effective way to make up for the shortage of petrochemical raw materials in China [20–22].

The reaction equation of acetylene carbonylation to acrylic acid is as follows:

$$HC\equiv CH + CO + H_2O \longrightarrow H_2C{=}\underset{H}{C}{-}\overset{O}{\overset{\|}{C}}{-}OH$$

Nickel carbonyl was initially used as a catalyst in the synthesis of acrylic acid by carbonylation of acetylene. Because of the high toxicity of nickel carbonyl, nickel salts of weak acids such as $NiBr_2$, nickel acetate and nickel naphthenate were used to replace nickel carbonyl, but the catalytic activity was significantly reduced. In general, a small amount of copper salt and acid were added to improve the activity, forming a composite catalyst system with nickel salt as the main catalyst, copper salt and acid as promotors, which had been widely used in industrial production of acrylic acid. However, the catalyst system had some disadvantages including carbon deposition, low selectivity to acrylic acid, and equipment corrosion.

The catalyst system was improved by using phosphine ligand. Numerous researches were carried out using triphenylphosphine (PPh_3) as a ligand in the catalyst system. The selectivity to acrylic acid was increased to 89.2% by a catalyst system of $NiBr_2$–$CuBr_2$–PPh_3 with molar ratio of 1:0.5:1.5. The catalyst performance was further improved by using tri-*tert*-butylphosphine ($P(t\text{-}Bu)_3$) as the ligand. In the presence of a catalyst system of $NiBr_2$–$CuBr_2$–$P(t\text{-}Bu)_3$–CH_3SO_3H with molar ratio of 1:0.5:1.5:0.5, acetylene underwent a carbonylation with carbon monoxide and water at 200 °C under 8.0 MPa for 1 h, resulting in a conversion of 75.8% based on acetylene and a selectivity of 98.6% to acrylic acid.

In order to ensure safe operation, acetylene was generally absorbed by a solvent and then the liquid phase is pressurized and introduced into the reactor. Different solvents used in the reaction system also affected the selectivity to acrylic acid. Tetrahydrofuran was usually used as the solvent in the carbonylation of acetylene.

The concentration of acrylic acid in the liquid product was about 20%; in addition, there were a large quantity of tetrahydrofuran solvent and small amounts of impurities. The tetrahydrofuran was recovered by distillation under reduced pressure. An azeotrope containing tetrahydrofuran of 94.4% and water of 5.6% was collected and recycled. After removal of tetrahydrofuran, the solution was subjected to extraction by using composite extractant toluene-butanone. Acrylic acid was obtained by fractional distillation of the extraction phase, and the extractant was recovered and recycled.

7.4.3 Uses of acrylic acid

Acrylic acid is mainly used for preparation of various esters such as methyl acrylate, ethyl acrylate, butyl acrylate and isooctyl acrylate, which are the comonomers for production of water-soluble coatings and adhesives. The second major use of acrylic acid is in the production of super absorbent polymer (SAP), which is one of the fastest growing acrylic products in recent years, and widely used in personal hygiene products, agriculture, forestry, horticulture and so on. Acrylic acid is also used in the production of sodium polyacrylate, which is used as a dispersant in circulating cooling water treatment and as a detergent builder in place of sodium tripolyphosphate. In addition, a small amount of acrylic acid is used for preparation of epoxy acrylate and acrylic adsorption resin.

In China, the use of acrylic acid for various esters accounts for 60% of the total consumption. SAP consumes 20% of the output of acrylic acid. The use of acrylic acid for detergent builder, special resins and water treatment agent accounts for 7.5%, 5.5% and 4.5%, respectively. Other use accounts for 2.5% of the total consumption.

7.5 Methyl acrylate

Methyl acrylate, also known as acrylic acid methyl ester, methoxycarbonylethylene, methyl propenoate, methyl 2-propenoate, methyl propenate, methyl prop-2-enoate, methyl ester of 2-propenoic acid and methyl ester acrylic acid, is a derivative of acetylene and carbon monoxide. CAS Registry Number: 96-33-3. Molecular formula: $C_4H_6O_2$, molecular weight: 86.0892. The chemical structure is

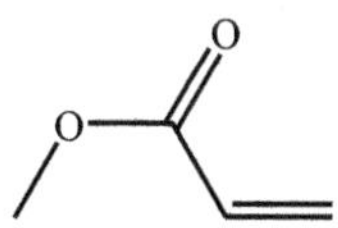

7.5.1 Properties of methyl acrylate

Methyl acrylate is a colorless liquid with a characteristic acrid odor. At 25 °C, the solubility of methyl acrylate in water is 5.2 wt% yet it decreases to 1–2 wt% in aqueous sodium chloride solution with a concentration of 20 wt%, while the solubility of water in methyl acrylate is 2.3%. Methyl acrylate is miscible with organic solvents such as methanol, ethanol and diethyl ether. It forms an azeotrope with water at 71.5 °C and 92.8% of methyl acrylate or 7.2% of water. The physical properties of methyl acrylate are listed in Table 7.5.

Table 7.5: Physical properties of methyl acrylate.

Melting point/°C	−75.7	Flash point/°C	10
Boiling point/°C	80.5	Enthalpy of fusion (−75.7 °C)/kJ · mol^{-1}	9.73
Vapor pressure (25 °C)/kPa	8.67	Heat capacity (25 °C)/J · mol^{-1} · K^{-1}	172.3
Density (20 °C)/g · cm^{-3}	0.9534	Enthalpy of vaporization (58 °C)/kJ · mol^{-1}	34.2
Refractive index (20 °C)	1.4040		

Methyl acrylate is prone to self-polymerization, so it should be stored in a dark environment below a temperature of 10 °C. It can be copolymerized with various monomers, and the higher the temperature, the more likely the copolymerization reaction occurs. Methyl acrylate is easy to react with strong acids, strong bases and strong oxidants, so it should avoid contact with such substances during storage. Methyl acrylate also softens materials such as plastics, rubber coatings and paints.

7.5.2 Processes for manufacture of methyl acrylate

Acetylene carbonylation, acrylonitrile hydrolysis and propylene oxidation are the three main processes for production of methyl acrylate in industry. The acrylonitrile hydrolysis process has been phased out due to that cumbersome disposal of by-product and the high cost of raw materials. The acetylene carbonylation process was used for large-scale industrial production in the 1960s and 1970s. In 1972, 60% of the total output of methyl acrylate in the United States came from the acetylene carbonylation process using acetylene as starting material. With the rapid development of the petroleum industry after the 1970s, the price of propylene was greatly reduced to one third of that of acetylene, and the cost of methyl acrylate by propylene oxidation process was 20% lower than that of acetylene carbonylation process. In addition, the pollution of acetylene carbonylation process was more serious. Therefore, the acetylene carbonylation process was gradually replaced by propylene oxidation process. At present, almost all methyl acrylate in the world is produced by propylene oxidation process. However, in recent years, with the rise of oil prices and the increase of oil imports in China, the cost of methyl acrylate by propylene oxidation has increased. Since China has abundant coal resources, and coal chemical industry is developing very rapidly, acetylene carbonylation has attracted much attention again.

7.5.2.1 Carbonylation of acetylene

As early as the 1940s, German scientist Reppe discovered that methyl acrylate could be prepared by carbonylation of acetylene with CO and methanol in the presence of nickel carbonyl as catalyst:

$$HC{\equiv}CH + CO + CH_3OH \longrightarrow H_2C{=}\underset{H}{C}{-}\overset{O}{\overset{\|}{C}}{-}O{-}CH_3$$

Nickel carbonyl and other metal carbonyl compounds were initially used as catalysts for the carbonylation of acetylene, but the metal carbonyl compounds were eliminated because of their high toxicity. Since then, nickel halide or nickel halide complexes have been used as the main catalyst, while copper halide and acid were used as the promotors.

7.5.2.1.1 Low-pressure Reppe process

Acetylene undergoes a carbonylation reaction with nickel carbonyl, water (or alcohol) and hydrochloric acid at a temperature of 40 °C and under atmospheric pressure to give methyl acrylate with a yield of 80%. The reaction equations are as follows:

$$HC{\equiv}CH + Ni(CO)_4 + 2HCl + 4H_2O \longrightarrow 4CH_2{=}CHCOOH + NiCl_2 + H_2$$

$$CH_2{=}CHCOOH + CH_3OH \longrightarrow CH_2{=}CHCOOCH_3 + H_2O$$

BASF had used this method to build industrial units. Although the reaction conditions are mild, the method has three obvious disadvantages: firstly, free hydrogen is released in the reaction, and a by-product propionate is produced, so that the yield of the acrylate is reduced, and the acrylate and the propionate are azeotropes, so that the product is not easy to separate; secondly, carbonyl nickel is very toxic, volatile, and large in use, it is difficult to achieve labor protection; and thirdly, $Ni(CO)_4$ needs to be regenerated from $NiCl_2$, and that regeneration process is troublesome. In view of the above, the method is difficult to realize industrial production in large scale.

7.5.2.1.2 Improved low-pressure Reppe process

After the reaction is started, carbon monoxide is introduced to replace the carbon monoxide in the nickel carbonyl, so that the recovery amount of nickel and the regeneration amount of the nickel carbonyl can be reduced. Most of the CO required for the reaction (80%) is provided by carbon monoxide gas, and only 20% comes from nickel carbonyl, so that the amount of nickel carbonyl is greatly reduced. The reaction temperature is 30–50 °C, the pressure is 0.1–0.2 MPa, and the molar ratio of methanol to total carbon monoxide is 1.1–3:1. The amount of hydrochloric acid must be maintained at 80–90% (mol) of nickel carbonyl in order to effectively inhibit the formation of by-product chloropropionic acid. The recovery of nickel and the regeneration of nickel carbonyl are as follows:

$$NiCl_2 + 6NH_3 \longrightarrow Ni(NH_3)_6Cl_2$$

$$Ni(NH_3)_6Cl_2 + 5CO + 2H_2O \longrightarrow Ni(CO)_4 + (NH_3)_2CO_3 + 2NH_3 + 2NH_3Cl$$

Rohm and Haas and Toagosei had used this method to produce methyl acrylate.

7.5.2.1.3 High-pressure Reppe method

Using tetrahydrofuran or acetone as a solvent, and in the presence of a catalyst system containing nickel halide as a main catalyst and copper halide and acid as cocatalysts, the carbonylation reaction was carried out at a temperature of 200–225 °C under a pressure of 8–10 MPa. The unreacted acetylene and carbon monoxide was recycled. The conversion of acetylene per pass was more than 90%, and the selectivity to methyl acrylate reached 85%.

Compared with nickel catalysts, palladium catalysts have high activities under mild conditions. In the presence of a catalyst system composed of palladium acetate, diphenyl-2-pyridylphosphine and *p*-toluenesulfonic acid, the carbonylation reaction was carried out at a temperature of 50 °C under a pressure of 1.0 MPa. The conversion of acetylene was 99.5% and selectivity to methyl acrylate was 99.7% under the optimal catalyst composition [23].

The advantage of this process is that the acetylene required for the reaction is dissolved in tetrahydrofuran in advance, which avoids the danger of gaseous acetylene under high pressure. Moreover, the catalyst used is nontoxic and replaces the toxic carbonyl nickel. The disadvantage is that the operation pressure is high, leading to the requirement of high standard of equipment.

7.5.2.2 Hydrolysis of acrylonitrile

In the early 1960s, due to the sharp decline in the price of acrylonitrile, a process for production of methyl acrylate from acrylonitrile was developed. Standard Oil Co. (Ohio) in the United States and Societe Ugine Company in France developed this process and realized industrialization, respectively. Some companies in Japan used this process for the production of methyl acrylate, and China also used this process to build a small-scale production unit.

The hydrolysis of acrylonitrile includes two steps. The first step is the hydrolysis of acrylonitrile to acrylamide in the presence of H_2SO_4 at 90 °C, and the second step is the esterification reaction of acrylamide with methanol at 150 °C in the presence of acid catalyst. Methyl acrylate is obtained by distillation under reduced pressure. It is necessary to add a polymerization inhibitor to prevent the polymerization of the acrylate, and the yield of acrylate is 90%. The reaction equations are as follows:

$$H_2C{=}CHCN + H_2O + H_2SO_4 \longrightarrow H_2C = CHCONH_3HSO_4$$

$$H_2C = CHCONH_3HSO_4 + CH_3OH \longrightarrow H_2C = CHCOOCH_3 + NH_4HSO_4$$

The process has the advantages include mild condition, high yield and easy operation. The disadvantage is that the by-product NH_4HSO_4 is difficult to handle, resulting in a large amount of waste liquid. This process has been replaced by propylene oxidation process, and now there is no industrial unit for production of methyl acrylate by hydrolysis of acrylonitrile.

7.5.2.3 Propylene oxidation process

Propylene is oxidized to acrylic acid by air in the presence of a catalyst. There are two ways: (1) direct oxidation of propylene to acrylic acid (one-stage oxidation process); and (2) propylene is first oxidized to acrylaldehyde and then further oxidized to acrylic acid (two-stage oxidation process). The yield of acrylic acid by one-stage oxidation process is only about 65% at most, while the yield of acrylic acid by two-stage oxidation process is about 80%. Therefore, two-stage oxidation process is currently applied in all industrial installations. Acrylic acid is then esterified with methanol to form methyl acrylate. The reaction equations are as follows:

$$H_2C = CHCH_3 + O_2 \longrightarrow H_2C = CHCHO$$

$$H_2C = CHCHO + O_2 \longrightarrow H_2C = CHCOOH$$

$$H_2C = CHCOOH + CH_3OH \longrightarrow H_2C = CHCOOCH_3$$

The performance of the catalyst directly determines the development level of the two-stage oxidation process. In the last decade, many companies in the world have modified the catalyst to improve the yield of acrylic acid, catalyst life and operating conditions. For example, the catalysts developed by Mitsubishi Chemical Corporation (MCC) and Nippon Shokubai (SKK) in Japan have the advantages of long service life and good activity, and are currently the most advanced catalysts. In China, Lanzhou Petrochemical Research Institute has developed different systems and series of catalysts such as LY-A-8801, LY-A-8802, LY-A-8201, LY-A-8202, LY-A-9601 and LY-A 9602. Shanghai Huayi Acrylic Acid Company has applied LY-A-9601 and LY-A-9602 catalysts in an industrial unit with an annual output of 30,000 tons, and the catalysts have also been used by Jilin Petrochemical Company.

Concentrated sulfuric acid is used as catalyst in the esterification reaction of acrylic acid with methanol in industry, which causes corrosion to equipment and side reactions. A number of new catalysts have been studied. The expandable graphite shows high catalytic activity in the esterification of acrylic acid to methyl acrylate, and the yield of methyl acrylate is more than 98%. The esterification catalyzed by the expandable graphite has the following advantages: short reaction time, no corrosion to equipment, no pollution and easy separation. MCC has also done research on *p*-toluene sulfonic acid catalyst, which greatly reduces the corrosion to equipment.

Since the 1980s, almost all the new and expanded industrial units in the world have adopted two-stage propylene oxidation process. At present, the companies having two-stage propylene oxidation technology include Sohio, MCC, SKK and BASF. Sohio technology had occupied a certain market in the late 1960s and early 1970s, but in the 1980s, technologies of MCC and SKK gradually replaced Sohio technology, exhibiting longer catalyst life, better selectivity and higher yield of acrylic acid.

In the propylene oxidation process, propylene needs to be oxidized to acrylic acid first, and then methyl acrylate is obtained via esterification reaction, while the acetylene carbonylation is a one-step reaction to directly obtain methyl acrylate, which is one step less than the propylene oxidation method. The Eighth Design Institute of the former Ministry of Chemical Industry of China made economic comparison of methyl acrylate production by acetylene carbonylation and propylene oxidation in 1989. The cost of methyl acrylate depended mainly on the price of the raw materials. According to the international price of raw materials at that time, the costs of methyl acrylate production by these two processes were almost the same, while according to the price of raw materials in China, the cost of methyl acrylate production by acetylene carbonylation process was 10% lower than that by propylene oxidation process. When the price of acetylene was three times that of propylene, the cost of methyl acrylate by acetylene carbonylation was 20% higher than that by propylene oxidation process; when the price of acetylene was twice that of propylene, the cost by the two processes was almost the same.

From the comparison of production costs, the acetylene carbonylation process was not inferior to the propylene oxidation process. Considering the basic situation of rich coal and poor oil in China, a non-petroleum approach for production of methyl acrylate is welcomed. However, it is necessary to improve the acetylene carbonylation process for a long life, no corrosive and eco-friendly catalyst system with higher selectivity.

7.5.3 Uses of methyl acrylate

Methyl acrylate is an important monomer for synthetic polymers. It is a soft monomer, which can be copolymerized with various hard monomers including styrene, acrylonitrile and vinyl acetate to obtain various polymer products. Methyl acrylate is mainly used as the second polymerization monomer for the synthesis of polyacrylonitrile fiber (acrylic fiber). Polyacrylonitrile fiber is one of the main synthetic fibers. The fiber with acrylonitrile content more than 85% is defined as polyacrylonitrile fiber, which is commonly known as artificial wool. However, pure acrylonitrile homopolymer has high glass transition temperature, and its dyeability and softness are not ideal, so other monomers need to be introduced for copolymerization modification. Modified monomers can be divided into two categories: one is called the second monomer, which is applied to reduce the glass transition temperature of polymer and improve fiber flexibility and mechanical properties. The most important second monomer is methyl acrylate, and its content in the copolymer is in the range from 5% to 10%. Another kind is called the third monomer, which is used to improve fiber dyeability and hygroscopicity. The most commonly used is olefin sulfonate, having a content of about 1% in the copolymer. The consumption of methyl acrylate on polyacrylonitrile accounts for more than 75% of the total output of methyl acrylate.

The remaining part of methyl acrylate is mostly used for copolymerization with various hard monomers such as styrene and vinyl acetate as well as functional monomers such as glycidyl ester, butyl acrylate, hydroxyethyl methacrylate, hydroxypropyl ester and their derivatives to synthesize some water-soluble, emulsion-type and solvent-type acrylic resin products. These products usually have low glass transition temperature, good weather resistance, good acid and alkali resistance and are widely used in leather and fabric processing, water treatment agents, adhesives and coatings.

In addition, methyl acrylate is also used in the synthesis of antibacterial and anti-inflammatory drugs in the pharmaceutical industry.

7.6 2-Methyl-1-acetylenyl-2-pentene-1-ol, empenthrin and chlorempenthrin

2-Methyl-1-acetylenyl-2-pentene-1-ol, also known as 1-ethynyl-2-methyl penten-2-ol, is a derivative of acetylene and 2-methyl-2-pentenal. Molecular formula: $C_8H_{12}O$, molecular weight: 124.18. The chemical structure is

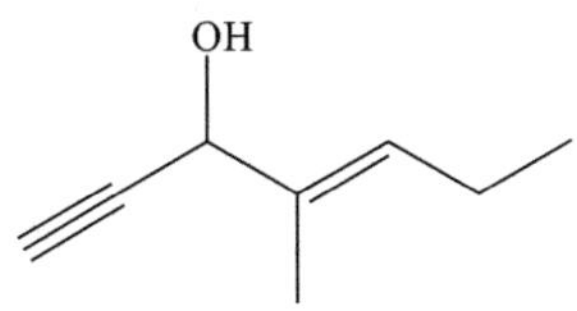

Empenthrin, also known as vaporthrin, (2*E*)-1-ethynyl-2-methyl-2-penten-1-yl 2,2-dimethyl-3-(2-methyl-1-propen-1-yl)cyclopropanecarboxylate, is a derivative of 1-ethynyl-2-methyl penten-2-ol. CAS Registry Number: 54406-48-3. Molecular formula: $C_{18}H_{26}O_2$, molecular weight: 274.404. The chemical structure is

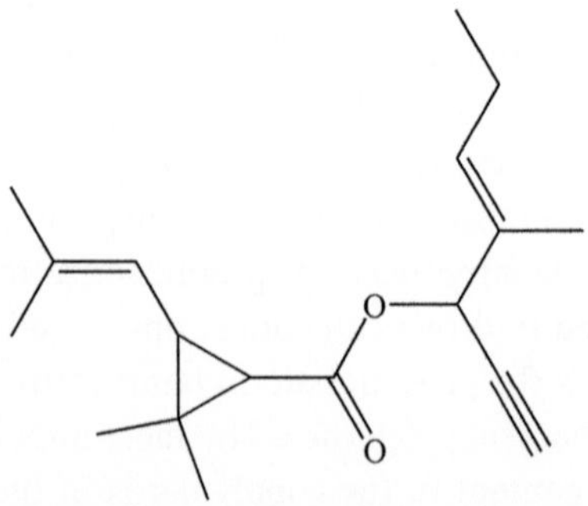

Chlorempenthrin, also known as 1-ethynyl-2-methylpent-2-enyl (*RS*)-2,2-dimethyl-3-(2,2-dichlorovinyl) cyclopropanecarboxylate and 3-(2,2-dichlorovinyl)-2,2-dimethylcyclopropanecarboxylic acid-1-ethynyl-2-methylpenten-2-ol ester, is a derivative of 1-ethynyl-2-methyl penten-2-ol. CAS Registry Number: 54407-47-5. Molecular formula: $C_{17}H_{22}O_2Cl_2$, molecular weight: 329.265. The chemical structure is

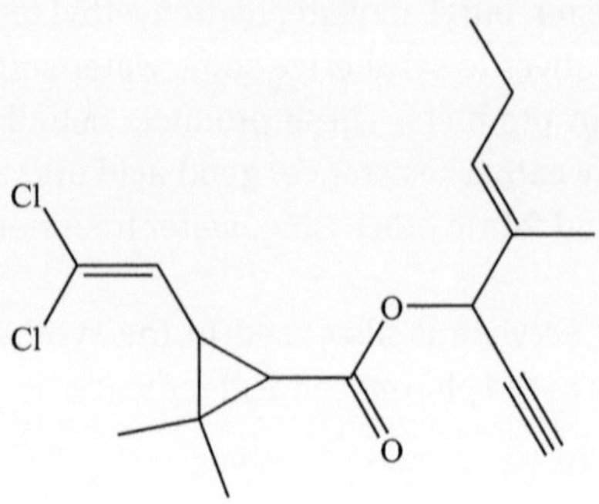

7.6.1 Properties of 2-methyl-1-acetylenyl-2-pentene-1-ol, empenthrin and chlorempenthrin

2-Methyl-1-acetylenyl-2-pentene-1-ol is a colorless and oily liquid with a reduced pressure boiling point of 112–115 °C at 10 kPa. It is insoluble in water but soluble in organic solvents such as ethers and alcohols.

Empenthrin is a light yellow and oily liquid with a boiling point of 346 °C and a flash point of 144 °C. Its density is 0.999 g cm^{-3} at 20 °C. The refractive index of empenthrin is 1.5270 at 20 °C. It has low acute mammalian toxicity (its oral LD50 is >5,000 mg kg^{-1} in male rats, >3,500 mg kg^{-1} in female rats and >3,500 mg kg^{-1} in mice) [24]. It is, however, very toxic to fish and other aquatic organisms.

Chlorempenthrin is a light yellow liquid or a light yellow to white solid with a reduced pressure boiling point of 128–130 °C at 40 Pa. Its density is 1.12 g cm^{-3} at 20 °C. The refractive index of chlorempenthrin is 1.5037 at 25 °C. It is easily soluble in organic solvents such as acetone, ethanol and benzene but insoluble in water.

7.6.2 Process for manufacture of 2-methyl-1-acetylenyl-2-pentene-1-ol, empenthrin and chlorempenthrin

7.6.2.1 Preparation of 2-methyl-1-acetylenyl-2-pentene-1-ol

Acetylene reacts with 2-methyl-2-pentenal to produce 2-methyl-1-acetylenyl-2-pentene-1-ol:

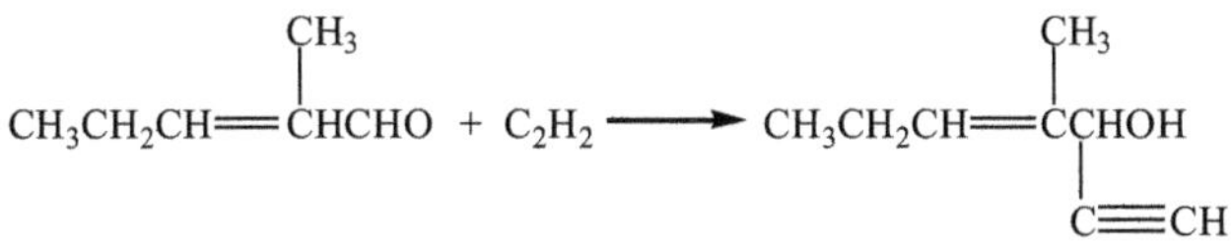

At a temperature of –50 °C, 3,000 mL of liquid ammonia was added to a 5,000 mL four-necked flask equipped with a stirrer, a thermometer, a dropping funnel and an outlet drying tube. Then 288 g (6 mol) of sodium amide was added to the liquid ammonia, and acetylene was introduced into the solution under agitation. About 594 g (6 mol) of 2-methyl-2-pentenal was added in a dropwise manner to the solution at a temperature of –35 to –50 °C. The reaction was continued for 4 h after the addition was completed. When the reaction is finished, an equimolar amount of ammonium chloride was added and the solution was stirred for 1 h. Subsequently, the temperature was elevated to evaporate the liquid ammonia for recycling. Water (1,500 mL) was added to the solution, and the solution was mixed vigorously with 1,500 mL of toluene and then stood for layering. The organic layer was recovered and washed with water until neutral. It was dried over sodium sulfate and then evaporated under reduced pressure. The fraction from 112 to 115 °C (10 kPa) was collected to obtain 640 g of the product with a purity more than 98% and a yield of 86%.

7.6.2.2 Preparation of empenthrin

About 3.72 g (0.03 mol) of 2-methyl-1-ethynyl-2-pentenol and 3.16 g (0.04 mol) of pyridine were added to toluene to form a solution. A toluene solution containing 5.6 g (0.03 mol) of Chrysanthemoyl chloride was added in a dropwise manner. The reaction was carried out at a temperature of 40–50 °C for 6 h. After reaction, the reaction solution was washed with water until neutral, dried and then subjected to distillation under reduced pressure. About 5.7 g of empenthrin was collected from a fraction of 120–126 °C (0.03 mmHg). The yield of empenthrin was 69% [25].

7.6.2.3 Preparation of chlorempenthrin

2-Methyl-1-ethynyl-2-pentenol, pyridine and toluene were added to the reaction vessel, and 3-(2,2-Dichlorovinyl)-2,2-dimethylcyclopropanecarbonyl chloride were added in a dropwise manner under agitation. The reaction was carried out at a suitable temperature for 6 h. After reaction, the reaction solution was washed with water until neutral, dried and then subjected to distillation under reduced pressure. Chlorempenthrin was collected from a fraction of 130–132 °C (1 mmHg). The yield of chlorempenthrin was 86% [26].

7.6.3 Uses of 2-methyl-1-acetylenyl-2-pentene-1-ol, empenthrin and chlorempenthrin

2-Methyl-1-acetylenyl-2-pentene-1-ol is a pesticide intermediate that can be used for manufacture of insecticides such as empenthrin and chlorempenthrin. These pyrethroids are repellents for mosquitoes, flies, cockroaches and clothing pests.

The vapor pressure of empenthrin at room temperature is suitable and therefore it has a natural volatility. At 30 °C, the vapor pressures of empenthrin and allethrin are 1.62×10^{-3} and 1.20×10^{-4} mmHg, respectively. Thus, the vapor pressure of empenthrin is about 10 times more than that of allethrin. Moreover, empenthrin is odorless. It has strong killing activity and repellent effect on pests and is very suitable for textile mothproofing. The mothproofing effect of two kinds of mothproofing agents that volatilized naturally on wool fabrics was compared in a glass vessel with a capacity of 700 mL. The mothproofing effect of empenthrin was more than 1,000 times higher than that of *p*-dichlorobenzene.

References

[1] Tachiyama T, Yoshida M, Aoyagi T, et al. Mechanistic study on dimerization of acetylene with a Nieuwland catalyst, Applied Organometallic Chemistry 2008, 22, 205–210.

[2] Tachiyama T, Yoshida M, Aoyagi T, et al. Deuterium Kinetic isotope effects and H/D exchange in dimerization of acetylene with a Nieuwland catalyst in aqueous media, Journal of Physical Organic Chemistry 2008, 21, 510–515.

[3] Tachiyama T, Yoshida M, Aoyagi T, et al. Deuterium and dimerization of acetylene with a Nieuwland catalyst in deuterium oxide, Chemistry Letters 2008, 37(1), 38–39.

[4] Tokita Y, Nishiwaki K, Kobayashi M, et al. Kinetics of copper(I)-catalyzed dimerization and hydration of acetylene in water, Bulletin of the Chemical Society of Japan 2004, 77, 1395–1399.

[5] Nishiwaki K, Kobayashi M, Takeuchi T, et al. Nieuwland catalysts: Investigation of structure in the solid state and in solution and performance in the dimerization of acetylene, Journal of Molecular Catalysis. A, Chemical 2001, 175, 73–81.

[6] Ohorodik A, Liblar E, Gehrmann K, et al. Production of monoviylacetylene. US 3957899. 1976

[7] Lu B L. Preparation of acetylene dimerization catalyst by one-step method, China Synthetic Rubber Industry 1983, 6(4), 267–268.

[8] Liu J, Han M, Wang Z. Effect of solvent on catalytic performance of anhydrous catalyst in acetylene dimerization to monovinylacetylene, Journal of Energy Chemistry 2013, 22, 599–604.

[9] Deng G, Wang Z, Liu G, et al. Study on nonaqueous catalyst for dimerization of acetylene, China Synthetic Rubber Industry 1991, 14(2), 106–108.

[10] Jiao S Y. Study on the optimum reaction conditions of dimerization of acetylene, Shanxi Chemical Industry 1993 (1), 13–16.

[11] Deng G, Li J, Chen R. Kinetics and mechanism of the dimerization of acetylene in the catalytic system of CuCl-KCl-HCl-H2O – relationship between reaction rate and acetylene partial pressure, China Synthetic Rubber Industry 1996, 19(2), 106–107.
[12] Ohorodik A, Gehrmann K, Legutke G, et al. Production of monoviylacetylene. US 3843743, 1974.
[13] Zhan F, Huang W, Zhu J, et al. External circulation bubble reactor for dimerization of acetylene to vinylacetylene, China Synthetic Rubber Industry 1995, 18(1), 15–18.
[14] Lynch M. Manufacture and use of chloroprene monomer, Chemico-Biological Interactions 2001, 135–136, 155–167.
[15] Cao M. Synthesis of dimethyl ethylene ethynyl methanol, Acta Polymerica Sinica 1958, 2(1), 63–65.
[16] Leitmannova E, Cerveny L. Leaf alcohol preparation, Perfumer & Flavorist 2008, 33, 50–59.
[17] Watson S C, Malpass D B, Yeargin G S. Preparation of unsaturated alcohols and intermediates therein. US 4,178,297, 1979.
[18] Zou H, Pan X, Lin G. Synthesis technology of 2-methylpyridine, Journal of South China University of Technology (Natural Science Edition) 2008, 36(11), 34–37.
[19] Boennemann H, Samson M Process for Preparation of 2-vinylpyridine from acetylene and acrylonitrile. US 4267329, 1981.
[20] Lin T J, Xie H, Meng X, Shi L. Characterization and catalytic application of Bi2O3/NiO composite oxides in the Reppe carbonylation to acrylic acid, Catalysis Communications 2015, 68, 88–92.
[21] Cui L, Zhou X, Yang X G. Effect of phosphine ligands on properties of $NiBr_2$-$CuBr_2$ catalyst for synthesis of acrylic acid by acetylene carbonylation, Chinese Journal of Inorganic Chemistry 2014, 30(7), 1600–1608.
[22] Cui L, Yang X, Zeng Y, Chen Y, Wang G. A unique nickel-base nitrogen-oxygen bidentate ligand catalyst for carbonylation of acetylene to acrylic acid, Molecular Catalysis 2019, 468, 57–61.
[23] Tang C M, Li X L, Wang G Y. A highly efficient catalyst for direct synthesis of methyl acrylate via methoxycarbonylation of acetylene, Korean Journal of Chemical and Engineering 2012, 29(12), 1700–1707.
[24] Haneko H, Kawasaki S, Misaki Y, et al. Mammalian toxicity of empenthrin (vaporthrin^R, S-2852F), The Journal of Toxicological Sciences 1992, 17, 313–334.
[25] Wang T, Zhang Y, Qian W, et al. Development and application of empenthrin, Medical Animal Control 1995, 11(1), 27–29.
[26] Zhang Y, Zhao X, Li J, et al. Synthesis and insecticidal effect of chlorempenthrin, Pesticides 1987, 1, 11–19.

Index

https://doi.org/10.1515/9783110714999-008

Bei Fragen zur Produktsicherheit wenden Sie sich bitte an:
If you have any questions regarding product safety,
please contact:

Walter de Gruyter GmbH
Genthiner Straße 13
10785 Berlin
productsafety@degruyterbrill.com